Electron Phonon Interactions — A Novel Semiclassical Approach

ELECTRON PHONON INTERACTIONS

A Novel Semiclassical Approach

ALBERT ROSE
Visiting Scientist
Chronar Corporation
Princeton, NJ
USA

World Scientific
Singapore • New Jersey • London • Hong Kong

Published by

World Scientific Publishing Co. Pte. Ltd.
P O Box 128, Farrer Road, Singapore 9128

USA office: World Scientific Publishing Co., Inc.
687 Hartwell Street, Teaneck, NJ 07666, USA

UK office: World Scientific Publishing Co. Pte. Ltd.
73 Lynton Mead, Totteridge, London N20 8DH, England

The publisher would like to thank the following publishers for their permission to reproduce the reprinted papers found in this volume:

American Institute of Physics (*Phys. Stat. Sol.*); General Electric Company (*RCA Review*); Physical Society of Japan (*J. Phys. Soc. Japan*); Plenum Publishing Corp. (*Physics of Disordered Materials* and *Disordered Semiconductors*).

To those who have not granted us permission before publication, we have taken the liberty to reproduce their articles without consent. We shall however acknowledge them in future editions of this work.

ELECTRON PHONON INTERACTIONS
— A NOVEL SEMICLASSICAL APPROACH

ISBN 9971-50-635-1

Printed in Singapore by JBW Printers & Binders Pte. Ltd.

To Lillian

Contents

Albert Rose: Papers on electron phonon interactions

Physical concepts in energy loss processes. *RCA Review*, **32** (1971) 463—488.

The acoustoelectric effects. *RCA Review*, **27** (1966) 98—139.

Energy losses by hot electrons. *RCA Review*, **27** (1966) 600—631.

Field and temperature dependance of electronic transport. *RCA Review*, **30** (1969) 435—474.

Energy losses by hot electronics in solids: A semiclassical approach. *Journal of the Physical Society of Japan*, Supplement, 21 (1966) 431—433.

Classical aspects of spontaneous emission. *Cooperative Phenomena*, ed. H. Haken and M. Wagner, Springer-Verlag, Berlin (1973) 303—307.

Spontaneous emission revisited. *Phys. Stat. Sol.* **61** (1980) 133—140.

A simple classical approach to mobility in amorphous materials (1). *Physics of Disordered Materials*, ed. D. Adler, H. Fritzsche and S. R. Ovshinsky, Plenum Publ. Corp., 1985, pp. 391—398.

A simple classical approach to mobility in amorphous materials (2). *Proc. of the Intl'l. Workshop on Amorphous Semiconductors*, World Scientific Publ. Co., 1987, pp. 29—34.

An extension of Einstein's treatment of spontaneous emission, *Disordered Semiconductors*, ed. M. A. Kastner, G. A. Thomas and S. R. Ovshinsky, Plenum Publ. Corp. 1987, pp. 389—394.

Chapter I

Overview

This monograph is a radical departure from the conventional quantum mechanical approach to electron-phonon interactions. It originated some years ago from an attempt to translate the customary quantum mechanical analysis of electron-phonon interactions carried out in Fourier space into a predominantly classical analysis carried out in real space. The result was far more successful than I had anticipated.

In broad outline, the "classical" approach achieved a remarkably simple model and formalism that unified a wide spectrum of physical phenomena. (Parenthetically, the term "Classical" is meant to include certain quantum mechanical constraints such as the quantization of energy which are obviously essential but which distract very little from the classical flavor of the argument. With this understanding we will in the interest of economy write "classical" as classical.

In the literature, the various electron-phonon interactions have been analyzed by a dozen different authors in different papers. Here they will all be analyzed by a single, relatively simple classical model. These electron-phonon interactions include polar and non-polar optical phonons, acoustic phonons that interact via deformation potential and via the piezo electric effect and phonons in metals.

The same simple model is shown to apply to electron interactions with the deep lying X-ray levels of atoms, with plasmons and with Cerenkov radiation.

The unifying concept that applies to all of these phenomena is a new definition of a coupling constant. The distinctive character of this coupling constant is that its values are for logical reasons confined to the range of zero to unity. The logical upper limit of unity provides a new insight into the phenomenon of electron self-trapping by phonons and into the phenomenon of spontaneous deformation of the conduction band (an early model of superconductivity due to H. Fröhlich).

The essentially classical interaction of an electron with its surround is clearly brought out to be the cause of spontaneous emission of phonons. The same concept of a classical interaction with its surround is shown to account also for the spontaneous emission of photons. The cause of spontaneous emission of photons has been the subject of controversy for the last seventy years ever since Einstein introduced the concept of spontaneous emission in 1917. The controversy revolved around the question of whether the cause of spontaneous emission is the classical phenomenon of radiation by an accelerated electron or a purely quantum mechanical effect due to the presence of zero point quanta. The controversy is resolved here in simple logical terms, namely, that Einstein's argument for the existence of spontaneous emission applies equally to phonons and to photons. In the case of phonons the cause of spontaneous emission is clearly the classical interaction of an electron with its surround. Even though the presence of zero point phonon quanta is an available cause, no one, to my knowledge has made use of it. It would be an artifical logic that would make use of zero point quanta for the emission of photons and, at the same time to ignore the presence of zero point quanta for the emission of phonons. It is curious that this argument has escaped notice for the last seventy years. Instead, the argument has usually been carried out in terms of a morass of the proper ordering of operators.

While the bulk of this monograph has to do with quanta of phonons and quanta of photons, we include a discussion of the acoustoelectric effect which is a purely classical phenomenon. The reason for this is to show that the coupling constant that we defined for use in modeling all of the electron-phonon interactions turns out to be valid also for the purely classical acoustoelectric effect. This universality of the newly defined coupling constant is one of the purposes of this monograph.

A most important aspect of the coupling constant is that it is equally applicable to amorphous and crystalline materials. To the extent that the dielectric constant is insensitive to whether a material is amorphous or crystalline the part of the mobility that is controlled by electron-phonon interactions is the same for the amorphous and crystalline state of a material. Experimental data by chemists on organic materials bear this out. Mobilities as high as 500 cm^2/volt-sec have been observed in amorphous materials such as liquid methane where elastic scattering is expected to be small. This is in strong contrast to a common association by many physicists that the amorphous state is inevitably associated with low mobilities around unity.

This application of our model to amorphous materials is particularly significant since, to my knowledge, an analytic formulation of mobility in amorphous materials using conventional arguments does not yet exist.

Much of the content of this monograph is taken from a series of papers published in the RCA Review:

Reprinted from *RCA Review*, **32** (1971) 463,

Chapter II

Physical concepts in energy loss processes

A. Rose, RCA Laboratories, Princeton, N. J.

Abstract—The loss of energy by an excited electron is known generally as spontaneous emission. The physical origin of this emission must logically lie in the interaction between the electron and its medium. The interaction in a solid medium, as opposed to vacuum, is easily traceable to a simple classical polarization of the medium by the emitting electron. A coupling constant is defined having the natural limits of zero and unity. It is shown that the constant can be expressed either in terms of the electrical and elastic components of the energy of the distorted medium or in terms of the real part of its dielectric constant. The structure of the imaginary part is shown to play a negligible role. The coupling constant is valid in the classical limit of acoustoelectric interactions as well as in the quantum limit of electron–phonon interactions. By detailed balance, it is also valid for induced as well as spontaneous emission. The form of the coupling constant gives an easy insight into several tangential problems; a generalized expression for the Lyddane–Sachs–Teller relation, the self-trapping of electrons, and the spontaneous deformation of a lattice.

Introduction

A major part of this series of papers* has been concerned with the rates of energy loss by fast-moving electrons in a solid medium. As the velocity of the electron is increased, the electron radiates energy first to acoustic phonons, then to optical phonons, impact ionizations, plasmons, x-ray levels, and Cerenkov radiation, in that order. The range of electron energies extends from 10^{-3} to over 10^5 electron volts.

* This is the concluding part of a series published in previous issues of RCA Review as follows: Part I, "Small Signal Acoustoelectric Effects," Vol. 27, p. 98, March 1966; Part II, "Rates of Energy Loss by Energetic Electrons," Vol. 27, p. 600, Dec. 1966; Part III, "Large Signal Acoustoelectric Effects," Vol. 28, p. 634, Dec. 1967; and Part IV, "Field and Temperature Dependence of Electronic Transport," Vol. 30, p. 435, Sept. 1969.

Each of these loss mechanisms has been treated separately in the literature,** which extends over the past sixty years. The modes of treatment cover a wide range of physical models as well as mathematical techniques. For the most part, the analyses are carried out in Fourier space as opposed to real space. While all of the energy losses must, in the strictest sense, be treated by quantum mechanical methods, the higher-energy losses to electronic excitations[1] and Cerenkov radiation[2] have also been examined or approximated by classical methods. The losses to phonons, on the other hand, have almost universally been treated by perturbation theory.

By contrast, we have attempted in this series to treat all of the loss mechanisms by a common model—a model that is relatively simple, graphic, essentially classical, and couched in real space. The necessary consistency with quantum principles has been obtained by imposing the more or less obvious constraints on the classical argument *after* the classical solution was arrived at. These constraints are (1) that the electron energy exceed the energy $\hbar\omega$ of the radiation it emits and (2) that the uncertainty radius of the electrons $\hbar/mv$ be less than the wavelength of emitted radiation. These constraints are sufficient to ensure agreement between the classical argument and the results of quantum mechanics for the *average* rate of loss of energy. The actual loss, of course, takes place via discrete quanta of energy and occurs stochastically in time and space.

The classical argument introduced a new coupling constant β whose values have the natural limits of zero and unity. β was defined as the fraction of the available coulomb energy of the electron that could be transferred to the medium. The coupling constant was successful not only in unifying the wide gamut of energy-loss mechanisms but also in relating the macroscopic acoustoelectric effects to the microscopic electron–phonon interactions. Because this form of coupling constant is new and because it has a broad significance extending beyond the field of energy loss, its meaning and evaluation are examined at length in this paper. The coupling constant bears, for example, on such problems as the Lyddane–Sachs–Teller relation, and the coupled phonon–photon luminescence emission by trapped electrons.

The present form of coupling constant, in particular the fact that it physically can not exceed unity, illuminates several diverse problems, namely, the range of validity of the concept of deformation potential; the concept, introduced by Frohlich,[3] of a spontaneous instability of the structure of energy bands caused by the presence of electrons; and

** See Chap. IV for list of references.

the concept, treated by Toyazawa,[4] of electrons self-trapped by acoustic phonons. Each of these items will be discussed.

Energy-loss phenomena are generally associated with the imaginary part of the dielectric constant. Our analysis, on the other hand, uses only the real part. The difference is more significant than the well-known free choice one has by virtue of the Kramers–Kronig relations to deal with either the real or imaginary parts of the dielectric constant. For fast electrons, the rate of emission of energy is actually independent of the form of the imaginary part of the dielectric constant, that is to say, independent of the magnitude of the damping constant γ. The reasons for this are discussed in the present paper.

Finally, we add a reminder that the classical approach to spontaneous emission relates this loss to an elementary and well-defined classical interaction between the electron and its medium and avoids whatever mystery may have been attached to it by the formalism of perturbation theory.

Origin of the Coupling Constant β

In the first of this series of papers a mechanical model was introduced as a prototype for the rates of loss of energy by fast electrons in solids. The mechanical model consisted of a stationary particle (analog of the electron) deforming some attendant mechanical system (analog of the solid medium in which an electron is immersed) so that an energy E_w, called an energy well, was stored in the mechanical system. The response time of the mechanical system was taken to be τ.

It is immediately evident that the maximum rate at which the particle can impart energy to the mechanical system is

$$\left.\frac{dE}{dt}\right|_{\max} \approx \frac{E_W}{\tau}. \qquad [1]$$

This is accomplished by letting the particle rest for a time τ sufficient to impart the energy E_W to the mechanical system; abruptly moving the particle to a new position so that the energy E_W is left in the wake of the particle; allowing the particle to remain at rest again for a time τ and again displacing it abruptly; and so on.

If the disturbance created by the stationary particle in the surrounding mechanical system has a diameter d, then the average velocity of the particle in the above argument is of the order:

$$v_o \approx \frac{d}{\tau}. \qquad [2]$$

If the particle now moves at a velocity v, considerably faster than v_o, it will traverse the diameter d in a time $T = d/v$ that is too short to form a complete energy well. Only a fraction $(T/\tau)^2$ of the complete energy well will be left in the wake of the moving particle. The reason for squaring the ratio T/τ is that the momentum imparted to the mechanical system is proportional. to T; hence the energy imparted will be proportional to T^2. In summary, then, the particle loses energy to the mechanical system at the rate:

$$\frac{dE}{dt} \approx E_W \left(\frac{T}{\tau}\right)^2 \frac{1}{T} \qquad \text{for } T < \tau$$

$$= E_W \frac{T}{\tau^2}. \qquad [3]$$

The parallel expression for the rate of loss of energy by an electron in a solid was obtained from Eq. [3] by using the following equivalents:

$$E_W = \beta \frac{e^2}{K_H d}, \qquad [4a]$$

$$T = \frac{v}{d}, \qquad [4b]$$

$$\tau = \omega^{-1}. \qquad [4c]$$

Eq. [3] then takes on the general form for rate of loss of energy by an electron;

$$\frac{dE}{dt} \approx \beta \frac{e^2 \omega^2}{K_H v}. \qquad [5]$$

We have omitted from Eq. [5] a geometric factor that comes from summing up Eq. [5] over a range of shell diameters d. The equivalents shown in Eq. [4a]-[4c] were derived by considering a fast electron losing energy to a mode (e.g., polar optical phonons) whose characteristic frequency is ω. The response time of the medium to an impulsive force is then, by definition, $\tau = \omega^{-1}$. If we fasten our attention on a particular spherical shell around the electron extending from about $d/2$ to $3d/2$, the disturbance in the medium also has these dimensions and the transit time is of order v/d. Finally, we recognize that the only way the electron can act on the medium (ignoring its spin) is via its coulomb field. The maximum energy the electron can impart to the medium is its vacuum coulomb energy e^2/d. If the electron is imparting energy by polarizing the medium, the maximum energy is reduced by K_H to $e^2/(K_H d)$ where K_H is the dielectric constant for frequencies higher than those under consideration (in the case of optical phonons, K_H is the electronic part of the dielectric constant). This part of the polarization permanently clings to the moving electron and masks the field and energy that it has available for doing work on the slower responding elements of the medium. The factor β (where $0 \leqslant \beta \leqslant 1$) was introduced purely formally to recognize that, in general, only a fraction of the available coulomb energy could be imparted to the medium. This fraction is not to be confused with the fraction $(T/\tau)^2$, which enters in explicitly to take into account the imperfect transfer of energy arising from the fast motion of the electron—fast compared with the response time of the medium. The factor β has to do with the fraction of coulomb energy that can be transferred to the medium when the electron is essentially stationary. That is, it may have a quantum mechanical kinetic energy of localization but it is not moving through the medium. The act of imparting energy is defined by the operation of suddenly displacing the electron in the medium. The energy left behind in the medium where the electron was is the energy imparted. This operational definition is, of course, designed to match the manner in which a moving electron leaves behind a trail of energy in the medium.

In addition to the formal definition of β as the fraction of available coulomb energy imparted to the medium, two other expressions for β were obtained for use in particular cases, namely,

$$\beta = \frac{K_L - K_H}{K_L}, \qquad [6]$$

and

$$\beta = \left.\frac{\text{Electrical energy}}{\text{Total energy}}\right|_{\text{Lattice deformation}} \qquad [7]$$

For well-separated resonances in a solid as, for example, the ionic resonances at a few hundredth of a volt and the electronic resonances at a few volts, the dielectric constants K_H and K_L in Eq. [6] refer to values above and below the resonance in question and on the flat parts of the dielectric constant versus frequency curves (see Fig. 2).

Eq. [7] is more general. If one thinks of any elastic deformation of a solid, there will be associated with this deformation an elastic energy and, in general, an electrical field energy. The total energy of the deformation will be the sum of the elastic energy and the accompanying electric field energy as, for example, in the case of a deformation in a piezoelectric solid. This is the total energy that appears in the denominator of Eq. [7]. The electrical energy that appears in the numerator denotes the energy that electrons can gain by relaxing to the new conditions of the deformed medium. This relaxation takes on a variety of forms. For example, a sine-wave deformation in a piezoelectric solid is accompanied by a sine-wave potential and field pattern. The electrical energy is the energy that electrons can gain by congregating in the troughs of the wave. In this case, the electrical energy is also equal to the electric field energy, i.e., $\mathcal{E}D/(8\pi)$. This is also true for polar optical phonons or for any deformation that creates a real, macroscopic electric field. We must note, parenthetically, that the electric fields that go to make up the elastic energy are, in contrast, atomic in dimensions and not coherent beyond an atomic layer. The term "real" is inserted as a reminder that the slope of an energy band caused by deformation or doping is not a real field in the Maxwellian sense even though it acts on free carriers as if it were. Note that, in the case of real, macroscopic electric fields, the field appears both as the numerator of Eq. [7] and as one term in the denominator. Hence, β can not exceed unity. This is consistent with its definition.

For deformations that are not accompanied by a real, macroscopic electric field, the total energy in the denominator of Eq. [7] is just the elastic energy. The numerator is still the energy that is gained when the free carriers relax into the new energy level system of the deformed medium.

The first part of this paper is concerned with the meaning of the coupling constant β. It will be derived both for mechanical and for electrical systems. It will include also a proof that the two expressions for β, Eqs. [6] and [7], are equivalent for deformations accompanied by real, macroscopic electric fields. The proof has the added bonus of yielding a generalized expression for the Lyddane–Sachs–Teller relation.[5] The meaning of the more general form for β, Eq. [7], will be discussed in relation to the question of whether β can exceed unity and in relation to the more conventional treatments of energy loss.

In summary, the central parameter determining the energy losses by electrons in solids is the coupling constant β. Eq. [7] gives its most general definition. β is a direct measure of the transfer of energy between two systems. And, finally, a reminder that the rates of energy loss can be used, as shown in Chap. V, to obtain the normal, temperature-dependent mobilities of electrons in solids.

Energy Transfer in Mechanical Systems

We derive here the fraction of the energy of a compressed spring that can be transferred to a second quiescent or undistorted spring. This will depend on the relative spring constants. It will turn out also to be a symmetric fraction so that it will be independent of whether the transfer is from first spring to second or vice versa. In both these respects the springs not only offer a graphic parallel for the transfer of energy in electrical systems but also yield the correct quantitative relations.

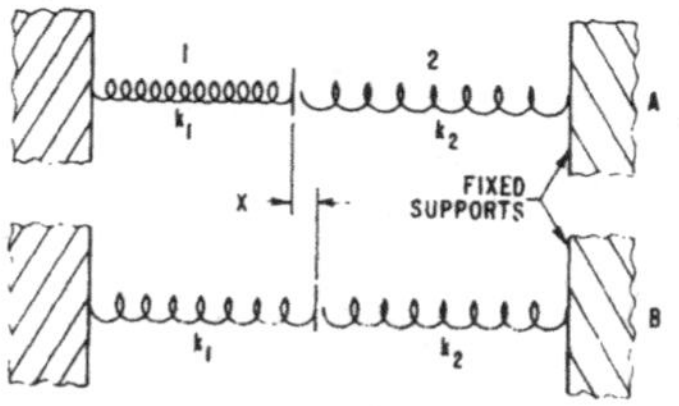

Fig. 1—Energy transfer between springs.

Consider, first, two springs with equal spring constants (Fig. 1). The first spring is compressed and the second undistorted. The two springs are in end to end contact. The first spring is released and the springs settle down after dissipating their kinetic energy so that they have equal amounts of stored energy. If the first spring was compressed by an amount s, the final rest position will obviously find each spring compressed by an amount $s/2$. This will mean that each spring has an elastic energy ¼ of the initial energy of the first spring since the elastic energy varies as s^2. In brief, ¼ of the initial energy of the first spring was transferred to the second. If we had examined the system before the kinetic energies were dissipated we would, of course, find equal amounts of kinetic and potential energy such that half the energy of the first spring had been transferred to the second spring. Equal spring constants represent the conditions for optimum transfer of energy.

We now examine the same problem but this time with different spring constant k_1 and k_2. Again, let the first spring be compressed an amount s. After release, the two springs will come to rest at a point at which their forces of compression are equal, namely,

$$k_1(s\text{-}x) = k_2x. \quad [8]$$

The energy transferred to the second spring is then

$$\Delta E = \tfrac{1}{2}k_2x^2, \quad [9]$$

and the ratio of this energy to the initial energy is

$$\frac{\Delta E}{E} = \frac{\tfrac{1}{2}k_2x^2}{\tfrac{1}{2}k_1s^2}. \quad [10]$$

From Eq. [8]

$$\frac{x}{s} = \frac{k_1}{k_1 + k_2}. \quad [11]$$

Insertion of Eq. [11] into Eq. [10] yields

$$\frac{\Delta E}{E} = \frac{k_1k_2}{(k_1 + k_2)^2}. \quad [12]$$

Eq. [12], by its symmetry, confirms the fact that the fractional energy transferred from one spring to the other is independent of which spring initially held the energy. Eq. [12] also reproduces the factor ¼ for optimum transfer when the spring constants are equal, or the factor of ½ if we add the kinetic energy.

Finally, for $k_1 \ll k_2$:

$$\frac{\Delta E}{E} \doteq \frac{k_1}{k_2}. \quad [13]$$

We can imagine, at this point, that k_1 represents an electrical energy (like that of a coulomb field) and that k_2 is the sum of two springs in parallel, one with the constant k_1 representing an electrical energy and the second with the constant k_3 representing an elastic energy. Then:

$$\frac{\Delta E}{E} = \frac{k_1}{k_2} = \frac{k_1}{k_1 + k_3} = \frac{\text{electrical energy}}{\text{electrical energy} + \text{elastic energy}}$$

$$= \left.\frac{\text{electrical energy}}{\text{total energy}}\right|_{\text{spring \#2.}} \quad [14]$$

If spring #2 represents the medium on which spring #1 (the electron) is doing work, we have confirmed, in brief, that the fraction of coulomb energy transferred to the medium is as shown in Eq. [7],

$$\beta = \frac{\Delta E}{E} = \left.\frac{\text{electrical}}{\text{total energy}}\right|_{\text{lattice deformation}} \quad [15]$$

Transfer of Energy in Electrical Systems

In this section, we will reproduce the spring arguments for an electrical system consisting of a capacitor with fixed charges and a dielectric that can be moved in and out of the capacitor. It will turn out that the energy that can be transferred from the electric field of the capacitor to the dielectric medium will follow the same pattern as that of the two springs. The arguments here, however, will be somewhat more involved algebraically but are worth going through in detail because they yield a generalized form for the Lyddane–Sachs–Teller relation. The capacitor with fixed charges acting on a removable dielectric is the equivalent of an electron moving through a polarizable medium.

Consider a medium whose dielectric constant has the classical form, with well-separated resonant frequencies ω_1, ω_2, etc;

$$K_o = 1 + \frac{\omega_{p1}{}^2}{\omega_1{}^2 - \omega^2} + \frac{\omega_{p2}{}^2}{\omega_2{}^2 - \omega^2} + \cdots, \quad [16]$$

where

$$\omega_{pi} = \frac{4\pi n_i c^2}{m_i},$$

and ω_i, n_i and m_i are the frequency, density, and mass, respectively, of each set of oscillators. The dielectric constant as a function of frequency is shown in Fig. 2. We have omitted from Eq. [16] and the figure the lossy (imaginary) term in the dielectric constant because we will be concerned with the dielectric constant only on the flat parts of the curve somewhat removed from the resonances. This simplifies the algebra. It also emphasizes the contrast between the present treatment, which computes electron energy losses using only the real part of the dielectric constant, and the more conventional treatments, which begin with the imaginary part of the dielectric constant. The contrast, as we will clarify later, is not simply the usual choice one has, by virtue of the Kramers–Kronig relations, of dealing with either the real or imaginary parts of the dielectric constant. It is more fundamental in that the dielectric loss mechanism itself has substantially no effect on the rate at which an electron loses energy to the medium.

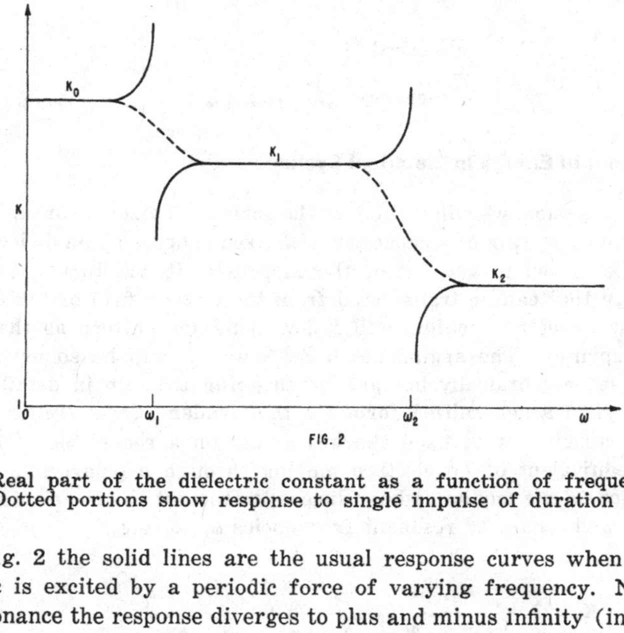

FIG. 2

Fig. 2—Real part of the dielectric constant as a function of frequency. Dotted portions show response to single impulse of duration ω^{-1}.

In Fig. 2 the solid lines are the usual response curves when the dielectric is excited by a periodic force of varying frequency. Near each resonance the response diverges to plus and minus infinity (in the absence of losses). If, on the other hand, we were to subject the dielectric to a single pulse of applied field, the width of the pulse being ω^{-1}, the response of the dielectric, that is, the amplitude of polarization (omitting small transient overshoots), would follow the dotted curve in passing through the various resonances. This is the type of field application we use below.

A further assumption is implicit in Fig. 2, namely, that the series of resonances make either comparable or increasing contributions to the dielectric constant but not decreasing contributions. The purpose of this assumption is to ensure that when we excite the ith resonance, the lower frequency resonances are negligibly excited. The assumption does not affect the conclusions drawn. The latter are based on the predominant excitation of a single resonance, independent of how the excitation is achieved.

Following Eq. [16], the dielectric constant K_i is

$$K_i = 1 + \frac{\omega_{pi}^2}{\omega_i^2 - \omega^2} + \frac{\omega_{pi+1}^2}{\omega_{i+1}^2 - \omega^2} + \cdots . \qquad [17]$$

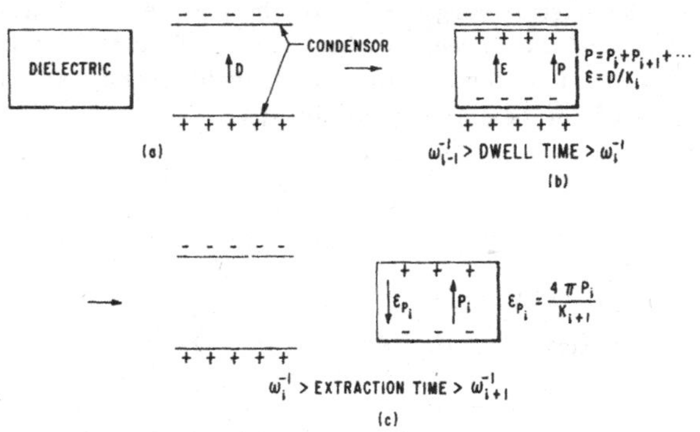

FIG. 3

Fig. 3—Polarization of a dielectric by transient exposure to an electric field.

Fig. 3 shows schematically the series of steps designed to impart the maximum energy to the ith mode of oscillation by transient exposure to an electric field. In brief, the dielectric is exposed to an electric field for a time long enough to polarize all of the oscillators beginning with ω_i and extending to higher frequencies but too short to polarize the lower frequencies. The dielectric is then removed in a time long enough to allow the oscillators with frequencies ω_{i+1} and higher to depolarize but too short to allow the oscillators ω_i to depolarize. In this way the oscillators ω_i are fully polarized by the electric field and retain that polarization after the field has departed. The energy of the polarized dielectric in this last step is the energy imparted to the medium.

The definition of β in terms of this model is the ratio of the energy of the polarized dielectric (Fig. 3C) to the "available" energy of the vacuum capacitor. The available energy of the capacitor is its vacuum field energy divided by K_{i+1}. The energy per unit volume of the polarized dielectric is

$$E_m \equiv \frac{1}{8\pi}\left[\frac{(4\pi P_i)^2}{K_{i+1}} + 4\pi\mathcal{E}P_i\right]. \qquad [18]$$

The first term is clearly the electric field energy of the polarization charge P_i. The second term is the elastic or "spring" energy that was stored in the polarized oscillators when they were first immersed in the capacitor field. To emphasize that this term is an elastic energy, and not an electrical energy in the sense of macroscopic electric fields, we can freeze the polarized dielectric and discharge the surface polarization charge so that no macroscopic fields remain. The remaining energy will then be the elastic energy $4\pi\mathcal{E}P_i$ of the polarized oscillators. Note that the energy of the dielectric while immersed in the capacitor field also contains the two terms—an electric field energy and the same elastic energy of the polarized oscillators;

$$\frac{\mathcal{E}D}{8\pi} = \frac{\mathcal{E}}{8\pi}(\mathcal{E} + 4\pi P_i + 4\pi P_{i+1} + \cdots) = \frac{K_i\mathcal{E}^2}{8\pi}$$

$$= \frac{1}{8\pi}(\mathcal{E}^2 + 4\pi\mathcal{E}P_i + 4\pi\mathcal{E}P_{i+1} + \cdots).$$

The distinction between electric field energy and elastic energy is also brought out below (see Eq. [25].

From Eq. [18] and the definition of β we write

$$\beta = \frac{\dfrac{1}{8\pi}\left[\dfrac{(4\pi P_i)^2}{K_{i+1}} + 4\pi\mathcal{E}P_i\right]}{\dfrac{1}{8\pi}\dfrac{D^2}{K_{i+1}}}, \qquad [19]$$

and use the conventional electrostatic relations,

and
$$\left.\begin{aligned} 4\pi P_i &= \mathcal{E}(K_i - K_{i+1}) \\ D &= K_i\mathcal{E} \end{aligned}\right\} \qquad [20]$$

to reduce Eq. [19] to

$$\beta = \frac{K_i - K_{i+1}}{K_i}. \qquad [21]$$

Hence, Eq. [6] is confirmed as a proper expression for β.

To confirm the equivalence of Eqs. [6] and [7] we write Eq. [7] in terms of the electric and total energy of the medium as given in Eq. [18];

$$\beta = \frac{\dfrac{1}{8\pi}\dfrac{(4\pi P_i)^2}{K_{i+1}}}{\dfrac{1}{8\pi}\left[\dfrac{(4\pi P_i)^2}{K_{i+1}} + 4\pi\mathcal{E}P_i\right]}, \qquad [22]$$

and use Eq. [20] to reduce Eq. [22] to

$$\beta = \frac{K_i - K_{i+1}}{K_i}. \qquad [23]$$

Hence Eqs. [6] and [7] are both valid for expressions for β.

The equivalence of Eqs. [6] and [7] can also be shown in a way that further illuminates the distinction between electrical and elastic energy.

From Eq. [16],

$$\frac{K_i - K_{i+1}}{K_{i+1}} = \frac{4\pi n_i e}{K_{i+1} m_i \omega_i^2}. \qquad [24]$$

Eq. [24] is valid on the flat parts of the dielectric constant versus ω (Fig. 2) where the dielectric constant is independent of ω. We now displace (polarize) each of the ith oscillators by an amount d so that a surface charge $n_i ed$ is formed. Then

$$\frac{K_i - K_{i+1}}{K_{i+1}} = \frac{4\pi n_i e}{K_{i+1} m_i \omega_i^2} \times \frac{4\pi n_i d^2}{4\pi n_i d^2}$$
$$= \frac{(4\pi n_i ed)^2}{8\pi K_{i+1}} \Bigg/ \tfrac{1}{2} n_i m_i (\omega_i d)^2 \qquad [25]$$

The numerator is the electrical energy owing to the surface charge $n_i ed$. The denominator is the maximum kinetic energy of the oscillators corresponding to an amplitude of oscillation d. The kinetic energy is also equal to the maximum potential energy of the oscillators—that is to say, their "spring" energy or elastic energy. Hence:

$$\frac{K_i - K_{i+1}}{K_{i+1}} = \frac{\text{Electrical Energy}}{\text{Elastic Energy}}, \qquad [26]$$

and

$$\frac{K_i - K_{i+1}}{K_i} = \frac{K_i - K_{i+1}}{K_{i+1} + (K_i - K_{i+1})}$$
$$= \frac{\text{Electrical Energy}}{(\text{Electrical} + \text{Elastic})\ \text{Energy}}$$
$$= \frac{\text{Electrical Energy}}{\text{Total Energy}}. \qquad [27]$$

Hence, the definitions of β given by Eqs. [6] and [7] are again shown to be equivalent.

Note that in Eq. [25], the electrical energy increases as the square of the density n of oscillators while the elastic energy increases only linearly.

The Lyddane–Sachs–Teller Relation

The free-standing polarized dielectric in Fig. 3C is an "electret," at least for the time required for the polarization to relax. In actual electrets the polarization is permanently frozen in by cooling and freezing a liquid dielectric in an electric field. The ionic part of the polarization is thereby frozen, while the electronic part is of course free to act. In actual electrets the ionic polarization is likely caused by the orientation of permanent dipoles and consequently does not carry with it an elastic energy.

The energy of the "electret" with its surface charge still intact is given by Eq. [18]

$$\text{Energy of "electret" (with surface charge)} = \frac{1}{8\pi}\left[\frac{(4\pi P_i)^2}{K_{i+1}} + 4\pi \mathcal{E} P_i\right]. \qquad [28]$$

With the aid of Eq. [20] this energy can be written in terms of D, the vacuum electric field of the capacitor used to induce the polarization in the "electret":

$$\text{Energy of "electret" (with surface charge)} = \frac{D^2}{8\pi}\left[\frac{1}{K_{i+1}} - \frac{1}{K_i}\right], \qquad [29]$$

If $K_i \gg K_{i+1}$ this energy reduces to

$$\frac{D^2}{8\pi K_{i+1}},$$

and, of course, if $K_{i+1} \to 1$ the energy is just that of the initial capacitor, as one would expect.

One can now discharge the surface charge of the "electret" (e.g., by ionizing the surrounding air) while retaining the frozen polarization. The remaining energy is then just the elastic energy given by the second term of Eq. [28];

$$\text{Energy of "electret" (without surface charge)} = \frac{\mathcal{E} P_i}{2},$$

or, with the aid of Eq. [20],

$$= \frac{D^2}{8\pi}\,\frac{K_i - K_{i+1}}{K_i^2}. \qquad [30]$$

This is the "frozen in" elastic energy that would be given up as heat if the "electret" were warmed to the point of releasing the polarization.

The Lyddane–Sachs–Teller relation is[7]

$$\frac{\omega_l^2}{\omega_t^2} = \frac{K_L}{K_H},$$

and relates the longitudinal and transverse frequencies of the polar optical modes of an ionic lattice to the low-frequency dielectric constant, K_L, and the high-frequency or electronic part of the dielectric constant, K_H. The squares of the frequencies are proportional to the energy per unit volume of the vibrating lattice. The transverse vibrations have only an elastic energy since macroscopic electric fields would violate the essentially conservative (or relatively static) character of the fields. The longitudinal vibrations have the same elastic energy plus an electric field energy due to the longitudinal polarization of the lattice. Hence, the Lyddane–Sachs–Teller relation can also be regarded as the ratio of the energies of a charged and a discharged "electret" keeping the volume polarization intact. This ratio from Eqs. [29] and [30] is

$$\frac{\text{Energy of charged "electret"}}{\text{Energy of discharged "electret"}} = \frac{K_i^2}{K_i K_{i+1}} = \frac{K_i}{K_{i+1}} = \frac{\omega_l^2}{\omega_t^2}. \qquad [31]$$

Eq. [31] is a generalized form of the Lyddane–Sachs–Teller relation holding even when there are more than the usual two resonances—ionic and electronic—of an ionic solid.

Coupled Phonon–Photon Emission

If a shallow trapped electron recombines with a free hole in the valence band to emit a photon, part of the energy is also radiated away in the form of phonons. Hopfield[6] computed the phonon energy to be

$$E_{\text{phonon}} = \frac{2}{(2\pi)^{1/2}} \frac{e^2}{\epsilon_\infty r} \left(\frac{\epsilon_o - \epsilon_\infty}{\epsilon_o} \right),$$

where $e^2/(\epsilon_\infty r)$ is the binding energy and r the radius of the shallow trapped electron. Williams[7] has confirmed this relation in studies of radiation from GaAs.

In terms of the definition of our coupling constant β, $e^2/(\epsilon_\infty r)$ is the available coulomb energy and $(\epsilon_o - \epsilon_\infty)/\epsilon_o$ is the expression for β for polar optical phonons. The product of these two factors is, then, the energy transferred from the coulomb field to the lattice in the form of elastic energy, and is the phonon energy left behind when the electron makes its radiative transition.

Range of Validity of Deformation Potential

The deformation potential B for acoustic phonons is defined[8] as the shift in energy of the bottom of the conduction band (or top of the valence band) per unit strain s of the lattice. By the nature of its definition, the concept cannot remain valid down to dimensions as small as a lattice spacing since deformations of one lattice spacing cannot be precisely associated with a band edge.

The present definition of coupling constant β offers another quantitative criterion for the dimension or wavelength at which the deformation potential no longer can remain a constant. The coupling constant β was defined as that fraction of the available coulomb energy that could be transferred to the lattice. Hence, β cannot logically exceed unity. In order to find the shortest wavelength for which B is a constant, we equate the β for acoustic phonons (see e.g. Part II of this series) to unity;

$$\beta = \frac{\pi K B^2}{C e^2 \lambda^2} \leqslant 1.$$

If we choose the following representative values: $K = 10$, $B = 4$ eV $= 6 \times 10^{-12}$ erg, $C = 10^{11}$ dynes/cm^2, e in e.s.u., then

$$\lambda \geqslant 2.5 \times 10^{-7} \text{ cm}.$$

For wavelengths shorter than 2.5×10^{-7} cm the value of B must decrease at least in proportion to the wavelength in order that β not exceed unity.

Lattice Instability

In an early discussion of a possible model for superconductivity, Fröhlich[3] proposed that, under certain conditions, the conduction band of a metal might spontaneously deform into a sine-wave pattern with a period of a few lattice spacings in order to achieve the lowest energy state for the metal. The periodicity of the conduction band edge would then give rise to a set of narrow forbidden gaps.

Conceptually, the spontaneous deformation of the lattice means that the electrical energy gained by the electrons settling into the troughs of the sine-wave distortion must exceed the total energy required to distort the lattice. Since the ratio of electrical to total energy of a lattice distortion is one of the ways of defining the coupling constant β, it must then follow that β would have to exceed unity. On the other hand, an equivalent definition of β as the fraction of available coulomb energy that can be imparted to the lattice precludes β exceeding unity. Hence, the lattice instability is logically excluded.

Self-Trapped Electrons

It is expected from the theory of polarons[9] that an electron can be self trapped in an ionic crystal provided

$$\alpha \equiv \frac{(\epsilon_o - \epsilon_\infty) e^2}{\epsilon_o \epsilon_\infty} \left(\frac{m}{\hbar^3 \omega} \right)^{1/2} > 6.$$

This condition is almost satisfied in ionic crystals and would be more than satisfied if the frequency ω of optical phonons could be substantially reduced from its usual value of about 10^{14}/sec.

Toyozawa[4] has examined the corresponding problem for non-ionic solids, namely, the criterion for self trapping of an electron by acoustic phonons and gives the condition

$$\frac{mB^2}{22\hbar^2 Ca} > 1,$$

where a is the dimension of the self-trapped electron. Its smallest value is a lattice spacing. We can explore the possibility of satisfying this criterion by comparing it with the expression for β;

$$\beta = \frac{\pi}{4} \frac{KB^2}{e^2 Ca^2} \leqslant 1,$$

where we have replaced λ by $2a$.

Noting from the earlier argument that B is a constant only for $\lambda = 2a \geqslant 2.5 \times 10^{-7}$ cm where the value of β approximates unity and that $B \propto \lambda$ for smaller λ, we can write

$$\frac{\pi}{4} \frac{KB^2}{e^2 Ca^2} = \left(\frac{2.5 \times 10^{-7}}{2a} \right)^2, \qquad a > 1.2 \times 10^{-7} \text{ cm},$$

or

$$\frac{B^2}{C} = \frac{4}{\pi}\frac{e^2}{K} \times 10^{-14}.$$

We insert this value into Toyozawa's condition to obtain

$$\frac{2 \times 10^{-14}}{11\pi} \frac{e^2 m}{\hbar^2 K a} > 1,$$

or, since $\hbar^2/e^2 m = 0.5 \times 10^{-8}$ cm,

$$\frac{10^{-6}}{8\,Ka} > 1, \qquad a > 1.2 \times 10^{-7}\ \text{cm}$$

The self-trapping condition is obviously not satisfied for dielectric constants $K > 1$, that is, for any real material.

For $a < 1.2 \times 10^{-7}$ cm, we give β its largest value, namely unity, and the above argument yields

$$\frac{10^8 a}{8K} > 1, \qquad a < 1.2 \times 10^{-7}\ \text{cm}$$

which, again, is not satisfied for real materials.

Real Versus Imaginary Part of the Dielectric Constant

When a dielectric is subjected to an ac field, the rate of loss of energy to the dielectric and its dissipation by the dielectric into heat is, by definition, given by the imaginary part of the dielectric constant. Hence, the conventional formalism for energy loss by an electron to a dielectric is couched (see Eq. [36], below) in terms of the imaginary part of the dielectric constant. The pulse of electric field exerted by the electron on an element of the dielectric must, of course, be resolved into its fourier components. Our formalism, on the other hand, is couched in terms only of the real part of the dielectric constant. While the Kramers–Krönig relations allow the imaginary part of the dielectric constant to be expressed in terms of the real part, one normally expects the damping constant γ (see Eq. [38]) to appear in the final expression. Our formalism does not contain γ. What we wish to show is that, in a profound sense, the fast-moving electron is insensitive to the presence or absence of dielectric loss processes.

In Fig. 4 we show an electron moving past three different types of elements that the electron might encounter in its medium. The first is a free charge, the second is a bound oscillating charge having negligible frictional loss, and the third is a bound oscillating charge having significant frictional loss. We assume, as is true for most of the range of

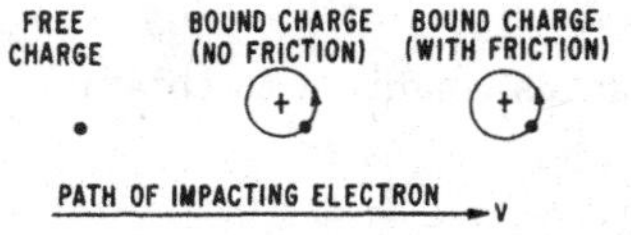

FIG. 4

Fig. 4—Three types of elements encountered by a high-velocity electron in a solid medium.

loss processes we have discussed, that the electron passes each of these elements in a time short compared with the response time of the element. That is to say, the electron gives each element an impulsive kick so that the element undergoes a vanishingly small displacement during the transit of the impacting electron. The meaning of this statement is that the impacting electron does not remain near each element long enough to distinguish whether the element is free, bound without friction or bound with friction. In each case, the impacting electron delivers substantially the same impulse (force $\times$ time) and, hence, the same energy, assuming the three elements to have the same mass. After the impacting electron has passed, each element disposes of its energy in its own way. The impacting electron is unaware of these subsequent events; it only knows that it has lost the same amount of energy to each of the elements. Since the impacting electron is unaware of the lossy nature of its medium, it is superfluous to introduce that information into the problem. Moreover, if it is introduced, it must then drop out in the subsequent analysis. It is for this reason that the rate of loss of energy by fast electrons can, at the outset, be expressed in terms of the real part of the dielectric constant only, and that substantially all trace of lossy nature of the medium be ignored.

The above arguments are valid for fast encounters in which the electron interacts with elements of the medium in times short compared with the response time of the element. When the converse is true, the electron does remain in the neighborhood of an element long enough to sense the lossy nature of the element. Here, the imaginary part of the dielectric constant does play a significant role. It should be noted, however, that, in the problems we have discussed, the rate of energy loss in this regime occurs at the maximum of the energy-loss-versus-electron-velocity curve and that this maximum can be approximated within a factor of two by a smooth extrapolation of the high-velocity part of the curve where only the real part of the dielectric constant is significant.

We show in the following argument how the conventional formalism, couched in terms of the imaginary part of the dielectric constant, can be converted, in the case of widely separated resonances, into a form containing only the real part of the dielectric constant and no trace of γ, the measure of the dielectric losses.

The rate of energy loss to polar optical phonons, is, in our formalism,

$$\frac{dE}{dt} = \frac{K_L - K_H}{K_L K_H} \frac{e^2 \omega_o^2}{v} \ln \frac{r_2}{r_1}, \qquad [32]$$

where ω_o is the optical phonon frequency, K_L the dielectric constant for frequencies below ω_o, and K_H the same for frequencies above ω_o. The energy loss takes place to elements of the medium lying between the radial distances r_2 and r_1 from the electron path. $r_2 = v\omega_o^{-1}$ and $r_1 = \hbar/(mv)$. If we fasten our attention on a particular radial shell surrounding the electron path such that the radius varies by only a factor of about 2, the logarithmic factor reduces to unity. From the classical form for the dielectric constant (see Eq. [16])

$$K_L - K_H = \frac{4\pi n e^2}{M \omega_o^2} \equiv \frac{\omega_p^2}{\omega_o^2}. \qquad [33]$$

Here n and M are the density and mass, respectively, of the ions that give rise to the optical phonon frequency ω_o. With these considerations, Eq. [32] becomes

$$\frac{dE}{dt} = \frac{1}{K_L K_H} \frac{e^2 \omega_p^2}{v}. \qquad [34]$$

If we assume further that $K_L - K_H \ll K_H$, then $K_L \approx K_H$ and Eq. 34 can be written

$$\frac{dE}{dt} \approx \frac{1}{K_H^2} \frac{e^2 \omega_p^2}{v}. \qquad [35]$$

We undertake, now, to show* how the conventional form for energy loss[10]

$$\frac{dE}{dt} = -\frac{2e^2}{\pi v}\int_0^\infty \frac{dq}{q}\int_0^\infty \omega I_m \frac{1}{K(q,\omega)}\, d\omega \qquad [36]$$

reduces to Eq. [35]. The integral over q in Eq. [36] leads to the logarithmic factor in Eq. [32]. By confining our attention to a shell in which the radius varies by about a factor of two, this integral yields a factor of unity. Moreover since we are concerned with the loss of energy to the polar optical phonons only, the upper limit of the integral over ω can be replaced by ω_o+, a frequency somewhat in excess of ω_o. Eq. [36] then becomes

$$\frac{dE}{dt} = -\frac{2e^2}{\pi v}\int_0^{\omega_o+} \omega I_m \frac{1}{K(\omega)}\, d\omega. \qquad [37]$$

We write K(ω) in the form

$$K(\omega) = 1 + \frac{\omega_p{}^2}{\omega^2 - \omega_o{}^2 - i\gamma\omega} + R, \qquad [38]$$

where R is the real part of the contribution to the dielectric constant of elements (e.g., valence-band electrons) having speeds of response much greater than ω_o. The imaginary part of these contributions will be substantially zero for frequencies near or below ω_o+ where the integration terminates.

From Eq. [38] it follows that

$$\mathrm{Im}\frac{1}{K(\omega)} = \frac{\omega_p{}^2\gamma\omega}{[(R+1)(\omega^2 - \omega_o{}^2) - \omega_p{}^2]^2 + (R+1)^2\gamma^2\omega^2}\,. \qquad [39]$$

By our assumption of $K_L - K_H \ll K_H$, $(R+1)\,\omega_o{}^2 \gg \omega_p{}^2$. It then follows, writing $\omega = \omega_o + \Delta\omega$, that the major contribution to the integral in Eq. [37] occurs near $\omega = \omega_o$ and in the range of $2\Delta\omega = \gamma$. Hence Eq. [37] becomes

---

* I am indebted to Dr. Smith Freeman for the ensuing argument.

$$\frac{dE}{dt} = \frac{2e^2\omega_p{}^2\omega\Delta\omega}{\pi v(R+1)^2\gamma\omega} = \frac{2}{\pi}\frac{e^2\omega_p{}^2}{(R+1)^2 v}$$
$$= \frac{2}{\pi}\frac{e^2\omega_p{}^2}{K_H{}^2 v}. \qquad [40]$$

Within these approximations Eq. [40] then matches Eq. [35]. The lossy nature of the dielectric, measured by γ, drops out in the evaluation of Eq. [36] because the integral has its major value in the range $2\Delta\omega = \gamma$. Another way of stating this is that Eq. [36] involves the integration of the imaginary part of the dielectric constant over a range of ω where it has a significant value, and the content of this integral is a constant independent of γ.

An alternative way of carrying out the argument is to evaluate $K_L - K_H$ through the Kramers–Krönig relation;

$$K_1(\omega) - K_\infty = \frac{2}{\pi}\int_0^\infty K_2(\mu)\,\frac{\mu}{\mu^2-\omega^2}\,d\mu,$$

where K_1 and K_2 are the real and imaginary parts of the dielectric constant. The result is in the form:

$$K_L - K_H \approx \frac{K_2(\omega_o)\Delta\omega}{\omega_o} = \frac{K_2(\omega_o)\gamma}{\omega_o} = \frac{\text{constant}}{\omega_o},$$

since $K_2(\omega_o)\gamma$ is a measure of the content of the imaginary part of the dielectric constant near ω_o and is a constant independent of γ.

Spontaneous Emission

We conclude with a reminder that the major part of it, the rates of energy loss by fast electrons in a solid, is concerned with spontaneous emission. In the literature on spontaneous emission in vacuum, the quantum mechanical formalism has been variously interpreted to reflect the perturbation of an excited state by zero-point vibrations in the vacuum field; to reflect the interaction between an excited state wave function and a possible mode of vibration in the vacuum; and, often, to reflect one of the mysteries peculiar to quantum mechanics, and not describable in classical language. Spontaneous emission in solids is subject to be the same formalism. In the solid, however, it is possible to trace the physical origin of spontaneous emission to the simple, classical, and graphic concept of the polarization of the solid by the electric field of the electron. The energy of the distorted solid left in the trail of the fast-moving electron constitutes the average rate of spontaneous emission. This classical concept leads to the definition of a coupling constant, β, which is equal to the ratio of electrical to total energy of the distorted solid (or medium). The coupling constant is symmetrical in the sense that it measures both the fraction of coulomb energy of the electron that can be transferred to the lattice and the fraction of distorted-lattice energy that can be transferred to electrons. The coupling constant is valid both in the classical limit of energy exchange with classical waves (acoustoelectric effect) and in the quantum limit of energy exchange with phonons (electron-phonon interactions). It is valid both for induced emission and for spontaneous emission processes, and indeed, is dictated by detail balance. Finally it is valid for amorphous as well as crystalline solids.

Acknowledgments

I am indebted to Professor M. A. Lampert and to Dr. A. Rothwarf for critical discussions of a number of parts of this paper.

References:

[1] N. Bohr, **Phil. Mag.**, Vol. 25, p. 10 (1913) and Vol. 30, p. 581 (1915)
[2] L. I. Schiff, **Quantum Mechanics**, p. 271, McGraw-Hill Book Co., New York, 1955.
[3] H. Fröhlich, "On the Theory of Superconductivity: the One-Dimensional Case," **Proc. Roy. Soc.** (London), Vol. A223, p. 296, 1954.
[4] Y. Toyazawa, **Polarons and Excitons**, p. 211, Oliver and Boyd, London (1963), ed. by C. G. Kuper and G. D. Whitfield.
[5] R. H. Lyddane, R. G. Sachs, and E. Teller, "On the Polar Vibrations of Alkali Halides," **Phys. Rev.**, Vol. 59, p. 673 (1941).
[6] J. J. Hopfield, "A Theory of Edge-Emission Phenomena in CdS, ZnS, and ZnO," **J. Phys. Chem. of Solids**, Vol. 10, p. 110 (1959).
[7] E. W. Williams, "A Photoluminescence Study of Acceptor Centres in Gallium Arsenide," **J. Appl. Phys.**, Vol. 18, p. 253 (1967).
[8] W. Schockley and J. Bardeen, "Energy Bands and Mobilities in Monatomic Semiconductors," **Phys. Rev.**, Vol. 77, p. 407 (1950).
[9] See, e.g., H. Fröhlich, **Polarons and Excitons**, p. 1, Oliver and Boyd, London (1963), ed. by C. G. Kuper and G. D. Whitfield.
[10] Taken from T. D. Schultz, **Quantum Field Theory and Many Body Problems**, p. 91, Gordon and Breach, New York.
[11] R. W. Smith, "Current Saturation in Piezoelectric Semiconductors," **Phys. Rev. Letters**, Vol. 9, p. 87 (1962).

Addendum to Part III

One of the notable and puzzling observations about the acoustoelectric effect is that the saturated drift velocities in GaAs and InSb are several times larger than the velocity of sound. In CdS and ZnO, the saturated drift velocities are, as expected, quite close to the velocity of sound. An argument was proposed in Part III to account for this difference in terms of the different behavior of the acoustoelectric effect in materials for which $ql > 1$ (GaAs and InSb) and those for which $ql < 1$ (CdS and ZnO).

A more likely source for this difference in behavior lies in the difference in coupling constants β for the two types of material. Qualitatively, one would expect that, in materials with a weak coupling constant, very high elastic strains would be needed to generate the electric fields required to bunch the carriers in the troughs of the acoustic waves. If this strain lies near the yield point of the material, the acoustic losses would be expected to increase so rapidly with increasing strain that the requisite strain could not be achieved.

Quantitatively, following the model initially cited by Smith,[11] one requires the electric field of the acoustic wave to be sufficient to accomodate the space charge of the electrons in a half wavelength. Thus:

$$\mathcal{E} = \frac{4\pi}{K} \cdot \frac{ne\lambda}{2}.$$

The value of $\lambda/2\pi$ at maximum gain is a Debye length:

$$\lambda^2 = (4\pi)^3 \frac{kT}{Kne^2}.$$

Also, by definition,

$$\beta = \frac{\text{Electrical energy}}{\text{Total energy}} = \frac{K\mathcal{E}^2}{4\pi CS^2}.$$

From these three equations, with $C = 5 \times 10^{11}$ dynes/cm², we compute the strain S required to trap all of the carriers:

$$S = \frac{5 \times 10^{-12}}{K} \left(\frac{n}{\beta}\right)^{1/2}$$

For CdS, $K = 9.3$ and $\beta = 5 \times 10^{-2}$. For GaAs, $K = 12.5$ and $\beta = 2.4 \times 10^{-3}$. These values give

$$S_{\text{CdS}} = 2.3 \times 10^{-12} n^{1/2},$$

and

$$S_{\text{GaAs}} = 8 \times 10^{-12} n^{1/2}.$$

Since the carrier densities are usually in the range of 10^{14}-$10^{16}/\text{cm}^3$, these expressions yield strains in the range of 10^{-4} to 10^{-3}, namely, in the range of plastic flow or brittle fracture. If, in GaAs, strains approaching brittle fracture are reached before all of the carriers are bunched in the troughs of the acoustic waves, it is likely that the acoustic losses will increase abruptly and the current will level off at values well above those expected for complete saturation at the drift velocity of sound. Evidence that the strains can be in the range of irreversible damage was found by Smith, in some of his early work on CdS, in the form of macroscopic holes drilled through crystals in which the acoustic flux was more than high enough to saturate the drift current.

Reprinted from *RCA Review*, 27 (1966) 98,

Chapter III

The acoustoelectric effects

BY

Albert Rose

RCA Laboratories
Princeton, N. J.

Editor's Note: This is the first part of a several part paper. Subsequent parts will appear in future issues of RCA Review.

In Part I the acoustoelectric effects for the several types of sound waves are obtained. The formalism used is the rate of exchange of energy between two coupled systems in relative motion. This formalism is shown to be consistent with an energy-well type of argument. The characteristic parameter in the final expressions is the ratio of electrical to total energy for the several types of sound waves.

In Part II the rates of loss of energy by free carriers is computed for loss to phonons, plasmons, electronic excitations and Cerenkov radiation. The formalism used is a classical energy-well argument upon which the quantum conditions are imposed as constraints. The characteristic parameter for loss to phonons is the ratio of electrical to total energy for the various forms of radiation.

In subsequent parts, the rates of energy loss of Part II are used to compute the mobilities of carriers in thermal equilibrium with the lattice, the mobilities of carriers whose mean energy is in excess of the temperature of the lattice, and the dependence of this mean energy on applied field. Finally, the conditions for dielectric breakdown via impact ionization by hot carriers are analyzed.

PREFACE

The content of this discussion emerged from an attempt to get a physical understanding of two areas of electron–phonon interactions — the low-field acoustoelectric current–voltage saturation first observed

* This work was supported in part by the U.S. Army Research Office–Durham, Durham, North Carolina under Contract No. DA-31-124-ARO (D)-84.

by R. W. Smith,[1] and the energy-loss mechanisms for hot electrons. The latter was actually to be a first step in interpreting the events leading to dielectric breakdown. It rapidly became clear that there was a closer connection between these two sets of phenomena than I had anticipated. In fact, this discussion argues that one can interpret a variety of phenomena, that are normally treated separately in the literature, by a single concept — that of an energy well — and by a single parameter — the ratio of electrical to total energy of an acoustic wave.

Throughout the course of this work I have had the generous guidance of a number of my colleagues in the extensive literature on electron–phonon interactions, and the concepts of polarons and plasmons. I have had, in particular, a steady exchange of ideas with M. A. Lampert, G. D. Whitfield (Pennsylvania State University), and R. H. Parmenter. Their contributions to the content of this work have been both substantial and substantive. In addition, I have profited on numerous occasions from discussions with L. R. Friedman, D. O. North, J. J. Quinn, and A. Rothwarf.

1. Introduction

AN ELECTRON in vacuum has a coulomb field extending from a radius at least as small as 10^{-12} cm to infinity. The energy density in the field is $\mathcal{E}_r^2/(8\pi)$, where $\mathcal{E}_r = e/r^2$. When the electron is plunged into a material medium, its electric field is reduced by the polarization of the medium to a value $\mathcal{E}_r/K$, where K is the relative dielectric constant of the medium. The polarization extends from some radius not smaller than one angstrom, owing to the atomic nature of any medium, toward infinity. The energy density in the reduced coulomb field in the medium is now $\mathcal{E}_r^2/(8\pi K)$. In brief, most of the coulomb energy of the electron* beyond some small radius has been polarized out. The potential energy of the electron has been reduced by the difference between $\mathcal{E}_r^2/(8\pi)$ and $\mathcal{E}_r^2/(8\pi K)$ integrated over the appropriate volume. The reduction in potential energy can be represented as an energy-well that the electron "digs" for itself in the medium.

If we now move the electron slowly through the medium, the polar-

---

[1] R. W. Smith, "Current Saturation in Piezoelectric Semiconductors," *Phys. Rev. Letters*, Vol. 9, p. 87, Aug. 1, 1962.

* The term "electron" is used for definiteness. The arguments apply equally to holes.

ization will follow the electron with negligible dissipation. As we speed up the motion of the electron, various components of the polarization will progressively be left behind as a "polarization wake" of the electron. The polarization arising from displacement of the atoms or ions will be the first to be abandoned owing to the relatively sluggish motion of atoms. The polarization arising from displacement of the electron clouds of the atoms will continue to follow the moving electron without significant dissipation until the electron has a velocity of several volts, at which time it too will tend to be left behind in the polarization wake of the moving electron. The polarization wake of the moving electron constitutes energy radiated by the electron in the form of phonons (i.e., lattice vibrations), plasmons (i.e., vibrations of the electron clouds), and finally, at velocities exceeding the velocity of light in the medium, photons (i.e., the electromagnetic radiation known as Cerenkov radiation). The energy-well that the "stationary" electron digs for itself by atomic displacements in the medium is the polaron well or polaron energy. For a medium with a single fixed frequency ω, the polaron well is a measure of the maximum rate of loss of energy by a moving electron. At sufficiently high velocities, the energy of the well left behind in unit *path length* by the moving electron decreases as the reciprocal of the square of the electron velocity. Hence the time rate of energy loss decreases as v^{-1}.

There is, of course, an extensive literature dealing with the excitation of phonons, plasmons, and photons by moving electrons. The treatments are, for the most part, quantum mechanical computations of the probability that an electron will make a transition between two states (of the electron) under the perturbing influence of thermal phonons, plasmons, or photons and their zero-point vibrations. The treatments, by their nature, must depart from elementary or pictorial concepts once the problem is formulated in terms of an integral over all of the possible quantum transitions.

Our purpose in this discussion is to try to match the formal and rigorous results of the quantum mechanical treatments by classical arguments upon which the constraints of the quantum nature of radiation and the wave nature of the electron are imposed in a way that retains the pictorial virtues of the classical concepts. It need scarcely be argued that such concepts are the working tools of most applied physicists. These concepts are especially helpful when one is faced with the results of a particular experiment, and sets about to inquire which of the manifold electron interactions is playing a significant role.

We will not be concerned in this discussion with numerical rigor and will be content in the interests of simplicity to match the more formal developments within a numerical factor of order unity. We *will* be concerned with *conceptual* rigor. Whether or not this is achieved, we hope that the examination of a collection of related phenomena in physical terms may facilitate the posing of conceptual questions by the non-theorist as well as by the theorist.

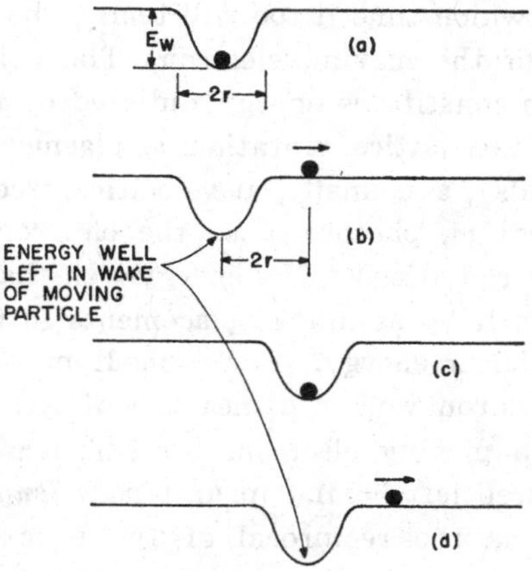

Fig. 1—Schematic steps showing maximum time rate of energy loss $(1/2)E_w\omega$ for a particle in an energy well.

2. Central Concept

The concept that underlies most of our discussion has already been referred to in the introduction. It is the idea that an electron digs an energy well by polarizing the surrounding material. When the electron moves, it tends to leave its well behind as a wake of radiated energy. We can make this concept more quantitative by a relatively simple estimate of an upper limit to the rate of loss of energy by a moving electron.

Let a "stationary" electron dig a well whose depth is E_w (see Figure 1). We put quotes on the term stationary because this classical concept must be reconciled with the uncertainty relation that sets limits on how precisely the position and momentum of an electron can be simultaneously specified. This comes up in later discussion. For the present we can imagine the electron sinking energetically into the medium much as a metal ball sinks into a rubber membrane. We specify the energy depth of the well by E_w, its physical extent by r, and the relaxation time of the medium by ω^{-1}. The relaxation time is derived, for example, from the ability of the medium to support a wave motion of angular frequency ω, or from the time required by a conductive medium to relax an electric field. It is the time required by the electron to dig its well by displacing the medium or by provoking currents that relax its electric field.

If we suddenly extract the electron from its well, we will leave behind an energy of polarization equal to the energy of the well. This operation has a familiar parallel in the physics of capacitors. Consider a pair of capacitor plates having a fixed charge. The charge on the capacitor plates is the analog of the electron in vacuum. We insert a material of high dielectric constant between the plates so that the energy of the capacitor is reduced from $(\mathcal{E}^2/8\pi)(Ad)$ to $(\mathcal{E}^2/8\pi K)(Ad)$. The product Ad is the volume of space between the capacitor plates. The insertion of the dielectric corresponds to plunging the electron from vacuum into a solid dielectric. Finally, we suddenly remove the dielectric from between the plates in a time short compared with any relaxation of charge in the dielectric. At the moment of removing the dielectric the increase of electrical energy of the dielectric owing to its polarization-induced surface charges will be $[\mathcal{E}^2/8\pi - \mathcal{E}^2/8\pi K]\,Ad$, or just the amount of energy by which the energy of the capacitor had been initially reduced by insertion of the dielectric. This last operation—removal of the dielectric—corresponds to suddenly separating the electron from its potential well.

Having suddenly extracted the electron from its well, we can locate it a distance $2r$, or one well-diameter removed, where it can proceed to dig a new well. The time required will, by our definitions, be ω^{-1}. After this time, we again displace the electron suddenly to a distance $2r$ removed. We make ω such displacements per second and at each displacement we leave behind (or radiate) an energy E_w. Hence, by this step-wise motion, we obtain an upper limit to the rate of loss of energy

$$\left.\frac{dE}{dt}\right|_{\max} = E_w\omega. \qquad (1)$$

This step-wise motion can obviously be replaced by a continuous motion such that the electron traverses the well diameter $2r$ in the relaxation time ω^{-1}. The effect on Equation (1) will be to multiply it by a factor of order 1/2, since the continuous as opposed to the step-wise motion will tend to leave perhaps half, but not all, of the energy of the well behind.

We anticipate parts of our later discussion by writing down here the maximum rate of loss of energy by an electron to polar optical phonons, and the maximum rate of loss of energy (per unit volume) by a sound wave* in a conducting medium.

Emission of polar optical phonons:

$$\left.\frac{dE}{dt}\right|_{max} = (\alpha_p \hbar\omega)\,\omega \tag{2}$$

where $\alpha_p \hbar\omega$ is the energy of a polaron, that is, the energy well† dug by a "stationary" electron in the polar optical modes of a crystal, and ω is the frequency of the optical modes—taken here to be a constant. α_p is the coupling constant in polaron theory. The use of the energy well is consistent with the operation, previously referred to, of integrating $\mathcal{E}_r^2 - (\mathcal{E}_r^2/K)$ over the appropriate volume.

Attenuation of sound waves:

$$\left.\frac{dE}{dt}\right|_{max} = \frac{1}{2} E_e \omega \tag{3}$$

where E_e is the density of electrical energy in the sound wave and ω is its frequency.

We note that Equation (2) for emission of polar optical phonons immediately matches Equation (1). In the case of Equation (3), we must recognize that E_e is physically or conceptually the energy well to be associated with an acoustic wave. That is to say, if we begin with an electrical energy density E_e in the sound wave we can reduce this energy to near zero by choosing the conductivity of the medium to have a dielectric relaxation time

$$\tau_r = \frac{K}{4\pi\sigma} = \omega^{-1}. \tag{4}$$

The electric field is then substantially polarized out everywhere. This operation is comparable to letting the electric field energy of an electron beyond a certain radius be polarized out by the medium. Hence, Equation (3) also matches Equation (1). Since Equation (3) is couched in

* The term "sound wave" is used to denote the traveling mode of lattice vibrations including both the acoustic and optical branches.

† The energy well is actually the polarization energy minus the kinetic energy of localization of an electron. This difference is about half the polarization energy itself.

terms of the electrical energy density of the sound wave, we will need to know the ratio of electrical to total energy in order to compute the attenuation of the sound wave itself, whose energy is predominantly elastic.

This ratio of electrical to total energy will play a central role in comparing the rates of loss of energy due to various phenomena. In fact, we list in Tables I and II (see pages 131 and 132) the several rates of loss of energy to be discussed in order to emphasize at the outset certain common features. These rates of loss of energy are either reproduced from the literature or are formal algebraic derivatives of the results as given in the literature. The common thread in all of these expressions is the ratio of electrical to total energy of the various wave phenomena considered. Tables I and II are useful in themselves as a convenient, common formalism for comparing rates of loss of energy due to the various phenomena. The tables also strongly suggest that there is a common set of physical concepts underlying the several phenomena. Our primary purpose is to explore these concepts.

To go beyond the central concept expressed in Equation (1), we need to compute the rates of energy loss for arbitrary velocities of the electrons relative to the wave phenomena with which they interact.

We begin with the several acoustoelectric effects because these phenomena are, for the most part, classical and present no conceptual difficulties. By introducing the ratio of electrical to total energy in a straightforward manner, the acoustoelectric effects offer an introduction to the phonon losses by individual electrons for which the same ratios of electrical to total energies appear as key parameters. As mentioned earlier, the relaxation by conduction currents of the electric field associated with the sound wave plays the same physical role as the relaxation by polarization of the electric field of individual electrons.

3. Acoustoelectric Effects

When a sound wave is propagated in a conducting crystal, the wave is attenuated owing to the I^2R losses incurred by the electric field associated with the wave. Parmenter[2] was the first to point out that the attenuation of the wave by free carriers should give rise to a force on the carriers tending to drag them in the direction of the wave. The result is a measurable acoustoelectric current or voltage. Weinreich[3]

---

[2] R. H. Parmenter, "The Acousto-Electric Effect," *Phys. Rev.*, Vol. 89, p. 990, March 1, 1953.

[3] G. Weinreich, "Ultrasonic Attenuation by Free Carriers in Germanium," *Phys. Rev.*, Vol. 107, p. 317, July 1, 1957.

re-examined the problem and gave a simple physical argument for expecting an acoustoelectric current to accompany the attenuation of the wave. The argument was that the work done on the wave represented a power that must be expressible as a force times a velocity. The velocity in this case was taken to be the wave velocity. The force must be the force of the wave on the free carriers and, hence, must reveal itself as a current or voltage.

Hutson and White[4] analyzed the attenuation of acoustic waves in piezoelectric materials where relatively strong electric fields accompany the wave motion. In a separate publication Hutson, McFee, and White[5] recognized and demonstrated that the attenuation of an acoustic wave could be symmetrically converted to an amplification of the wave by causing the electrons to drift faster than the wave and in the same direction. Hutson's analysis was a small-signal analysis that combined Maxwell's equations, the piezoelectric equations, and carrier diffusion effects into a single analytic solution.

Gunn[6] carried out an independent analysis of the amplification of polar optical sound waves by a drifting stream of carriers. Gunn's result matches that of Hutson when the ratio of electrical to total energy appropriate to polar optical waves is used to replace that appropriate to piezoelectric waves. Woodruff[7] carried out a more detailed analysis of the same problem.

A number of other somewhat more formal and more general analyses of the amplification of sound waves by Spector[8] and by Eckstein[9] have appeared.

We will first derive Hutson's result, ignoring diffusion, by making use of a relatively simple mechanical argument. The effect of diffusion will then be added by physical reasoning.

---

[4] A. R. Hutson and D. L. White, "Elastic Wave Propagation in Piezoelectric Semiconductors," *Jour. Appl. Phys.*, Vol. 33, p. 40, Jan. 1962.

[5] A. R. Hutson, J. H. McFee, and D. L. White, "Ultrasonic Amplification in CdS," *Phys. Rev. Letters*, Vol. 7, p. 237, Sept. 15, 1961.

[6] J. B. Gunn, "Travelling-Wave Interaction between the Optical Modes of a Polar Lattice and a Stream of Charge Carriers," *Physics Letters*, Vol. 4, p. 194, 1 April 1963.

[7] T. O. Woodruff, "Interaction of Waves of Current and Polarization," *Phys. Rev.*, Vol. 132, p. 679, 15 Oct. 1963.

[8] H. N. Spector, "Amplification of Acoustic Waves through Interaction with Conduction Electrons," *Phys. Rev.*, Vol. 127, p. 1084, Aug. 15, 1962; "Ultrasonic Amplification in Extrinsic Semiconductors," Vol. 130, p. 910, 1 May 1963; "Quantum Effects in the Amplification of Sound in the Presence of a Magnetic Field," Vol. 132, p. 522, 15 Oct. 1963.

[9] S. Eckstein, "Resonant Amplification of Sound by Conduction Electrons," *Phys. Rev.*, Vol. 131, p. 1087, 1 Aug. 1963.

3-A—Energy Exchange between Moving Systems

Consider two systems A and B in relative motion and let there be a force of interaction F between the systems. For definiteness, one may imagine (see Figure 2) a block A sliding on a block B and subject to a force of friction F. For convenience, let B rest (or slide) on a frictionless surface.

We move A at a constant velocity v_A such that it slides on B and exerts a force F on B. In this example, B will be uniformly accelerated.

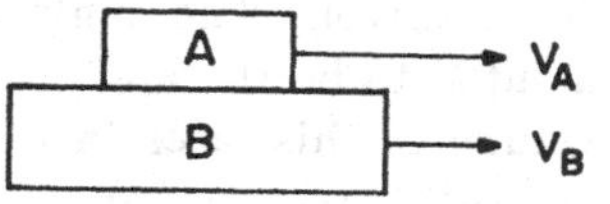

Fig. 2—Schematic diagram of body A sliding on body B.

However, we examine the power delivered to A and B at some particular velocity of B, namely $v_B < v_A$. We can write by inspection:

$$\left.\begin{array}{l}\text{power expended on A}\\ \text{by agency that moves A}\end{array}\right] \equiv P_A = F v_A, \tag{5}$$

$$\left.\begin{array}{l}\text{power expended on B}\\ \text{to increase momentum of B}\end{array}\right] \equiv P_B = F v_B, \tag{6}$$

$$\left.\begin{array}{l}\text{power dissipated into}\\ \text{"zero momentum" processes}\end{array}\right] \equiv P_D = F (v_A - v_B). \tag{7}$$

The velocities v_A and v_B are measured relative to a stationary observer. Since A by assumption is not accelerated, the power expended on A must be passed on *in toto* to increasing the momentum of B and to "dissipative" processes. This equality is evident by inspection of Equations (5)-(7).

The meaning of Equation (7) is perhaps not so obvious and needs a word of explanation. We have stated that the power input P_A to A must be passed on *in toto* to increasing the momentum of B and to other processes. Since P_B, the power used in increasing the momentum of B, is clearly less than P_A, there must, by conservation of energy, be a residual power, P_D, to make up the difference. By Newton's laws, this power cannot go to increasing the net momentum of the system and hence must go into processes whose net momentum is zero. Such processes in the case of a block A sliding on a block B are clearly the frictional heating of A and B at their surfaces. In a more general case, P_D can be associated with microscopic parts of B that are scattered symmetrically to either side of the line of flight of A. The sidewise momenta of such parts would then add up to zero, but their energies would of course be finite. A mechanical illustration of this is the splitting by an axe of a block of wood hung from a string. There will then be two parts to the energy of the pieces of wood—that associated with the net forward momentum of the pieces along the line of flight of the axe and that associated with the lateral components of velocity of the pieces whose momenta sum to zero.

An illustration more pertinent to our discussion is the Cerenkov radiation emitted by a fast electron. The angle at which the radiation is emitted is readily computed to be the angle that satisfies Equations (6) and (7). We will return to this later in the discussion.

Equations (5)-(7) are quite elementary. They nevertheless permit a radical simplification in the computation of the gain or attenuation of sound waves by a drifting stream of carriers. In the acoustic problem we are given the velocity v_A corresponding to the drift velocity of the carriers, and we are given also the velocity v_B corresponding to the phase velocity of the sound wave. What we need to know in order to compute P_B, the power going into the acoustic wave, is the force F. But from Equation (7) we can compute F providing we have a means of computing P_D, the dissipative power. This turns out, as we shall show shortly, to be an elementary computation for the relative motion of sound waves and a drifting stream of carriers. Hence, we have a simple formalism for computing the rate of doing work (Equation (6)) on the sound wave.

3-B—Amplification and Attenuation of Sound Waves

Consider an acoustic wave having associated with it an electric field $\mathcal{E}$ that undergoes the same sinusoidal variation in time and space as does the sound wave. If the sound wave is immersed in a medium of free carriers and if the carriers are drifting with a velocity equal to the phase velocity of the wave, there will be no steady exchange of energy between the wave and the carriers. The carriers will, by ohmic conduction, relax the electric field of the wave much as if the problem were one of a stationary electric field in a stationary conducting medium. Once the relaxed distribution of carriers has been established, the pattern of carrier distribution will, by our assumption of equal velocities, keep step with the electric field pattern of the wave. For example, electrons bunched in the positive troughs of the wave will remain in the same positive troughs as they move along with the phase velocity of the wave. Under these circumstances, there will be zero exchange of energy between the two systems.

We should note parenthetically that while the drifting electrons do no work on the particular sound wave we have chosen, they *do* transmit energy to the system of phonons of the solid. This is the I^2R loss that exists due to a current I whether the particular sound wave we are considering is present or not. We ignore or subtract out this source of dissipation as being irrelevant to the problem at hand. That is, the two systems we are considering are the drifting electrons and a particular sound wave.

At this point we give the electrons a drift velocity larger than the velocity of the sound wave. Now the system of free electrons, instead of seeing a static field as they did when they kept step with the wave, will see an alternating electric field whose apparent frequency is given by

$$\omega_r = \frac{v - v_s}{v_s}\,\omega. \tag{8}$$

Here, v is the drift velocity of the electrons, v_s the phase velocity of the sound wave and ω its frequency. The alternating electric field will contribute a dissipative power over and above that necessary to merely give the electrons a drift velocity v. It is this added dissipation arising from the presence of a sound wave that we wish to compute.

The dissipation per unit volume in a medium of conductivity σ subjected to an a-c field is an elementary problem whose solution is

$$P_D = \frac{\mathcal{E}_0{}^2\sigma}{2}\,\frac{1}{1 + (\omega_c/\omega_r)^2}$$

$$= \frac{\mathcal{E}_0{}^2\sigma}{2}\,\frac{(\omega_r/\omega_c)^2}{1 + (\omega_r/\omega_c)^2}, \tag{9}$$

where $\omega_c = 4\pi\sigma/K$, K is the relative dielectric constant, and $\mathcal{E}_0$ is the peak a-c field strength.

The physical problem and the equivalent circuit problem are shown in Figure 3. The physical problem consists of an a-c charge across a conducting dielectric while the equivalent circuit consists of an a-c voltage across a resistance in series with a capacitor.

From Equations (6)-(9) we compute the power delivered to the wave by the drifting stream of carriers;

$$\frac{dE}{dt} \equiv P_B = P_D \frac{v_s}{v - v_s} = P_D \frac{\omega}{\omega_r}$$

$$\frac{dE}{dt} = \frac{\mathcal{E}_0^2 K}{8\pi} \omega \frac{\omega_r/\omega_c}{1 + (\omega_r/\omega_c)^2}. \tag{10}$$

Equation (10) matches Hutson's result if we ignore his diffusion term. To make the match with Hutson more evident we compute the gain constant ω_a defined by

$$E_{tot} = E_{tot}|_0 \exp \omega_a t, \tag{11}$$

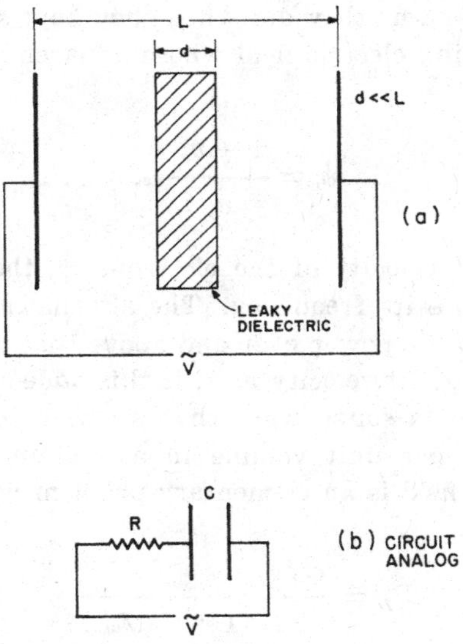

Fig. 3—Physical and circuit analogs of dissipation due to relative motion of the a-c electric field of a sound wave and a relaxable medium.

from which

$$\frac{dE_{tot}}{dt} = \omega_a E_{tot},$$

and

$$\omega_a = \frac{1}{E_{tot}} \frac{dE_{tot}}{dt}. \tag{12}$$

Here E_{tot} is the total energy density of the sound wave which is also equal to the kinetic or mechanical energy of the atoms at the time of zero displacement from their normal sites. That is, E_{tot} is the peak value of mechanical energy. It can be written in this form.

$$E_{tot} = \frac{1}{\beta} E_e = \frac{1}{\beta} \frac{\mathcal{E}_0{}^2 K}{8\pi}, \tag{13}$$

$$\text{where } \beta = \frac{\text{peak electrical energy}}{\text{total energy}} = \frac{\text{peak electrical energy}}{\text{peak mechanical energy}}. \tag{14}$$

In Equation (12) we replace E_{tot} by the right-hand side of Equation (13) and dE_{tot}/dt by the right-hand side of Equation (10) to obtain

$$\omega_a = \beta\omega \frac{\omega_r/\omega_c}{1 + (\omega_r/\omega_c)^2}. \tag{15}$$

Equation (15) is a particularly convenient form for quantitative estimates. For example, if $\beta \approx 1$ and $\omega_r = \omega_c$, the maximum rate of gain (or attenuation) is achieved, and corresponds to doubling (or halving) the energy of the sound wave at a rate of half the angular frequency of the wave. For $\beta < 1$, the maximum rate at which the wave gains energy would, according to Equation (10), be enough to double the *electrical* energy at this same rate of $\omega/2$. However, since the electrical energy is only a fraction of the total energy, and since the *latter* must be doubled to double the electrical energy, the rate of growth of either energy, that is, of the wave itself is reduced to $\beta\omega/2$.

We note also from Equation (10) that the maximum rate of change of energy of the sound wave is, as pointed out in section 2,

$$\left.\frac{dE}{dt}\right|_{\max} = \frac{1}{2} E_e \omega, \tag{16}$$

and occurs at $\omega_c = \omega_r$, that is, when the velocities of electrons and wave are such that their relative displacement is $\lambda/(2\pi)$ in the dielectric relaxation time $\omega_c{}^{-1}$. This is the concept, already cited, that the critical relative velocity of two systems needed to produce a maximum rate of transfer of energy is given by the diameter of the energy well divided by the relaxation time of the medium.

At low relative velocities ($\omega_r \ll \omega_c$) the charge pattern moves adiabatically with the wave with vanishing dissipation and vanishing work done on the wave. This is a major factor responsible for the small attenuation of sound waves in metals for which $\omega_c \gg \omega_r$.

At high relative velocities ($\omega_r \gg \omega_c$) the electrons drift a distance of one wavelength in a time too short to relax the fields of the sound wave. Hence, the work done on the wave again approaches zero. In this regime Equation (10) can be written to compare with Equation (1),

$$\frac{dE}{dt} = \left(E_e \frac{v_s}{v} \right) \omega_c, \tag{17}$$

where the depth of energy well is $E_e v_s / v$ and the relaxation time of the medium is ω_c^{-1}. The depth of well decreases as v^{-1}, as cited in the introduction.

3-C—Effect of Diffusion

Equation (15) for the gain constant was derived by computing the dissipation in a conducting medium due to an a-c field. Normally, one assumes that the density of carriers is not significantly perturbed in the process of relaxing the applied field. This is a valid assumption for long wavelengths of the applied field pattern. For sufficiently short wavelengths, however, the perturbed density of carriers can give rise to significant diffusion currents that tend to oppose the complete relaxation of the applied field. M. A. Lampert has shown (see Appendix I) that the inclusion of the diffusion term in the dissipation argument used to derive Equation (15) yields the complete Hutson result;

$$\omega_a = \beta \omega \frac{\omega_r / \omega_c}{(1 + \omega_D / \omega_c)^2 + (\omega_r / \omega_c)^2}, \tag{18}$$

where

$$\omega_D = \frac{4\pi^2 k T \mu}{e \lambda^2}. \tag{19}$$

Note that our definition of ω_D is the reciprocal of the diffusion transit time of an electron across $\lambda/(2\pi)$ and is not to be confused with Hutson's ω_D, which he defined as $v_s^2 e/(kT\mu)$.

There are several ways of interpreting the diffusion term ω_D/ω_c in Equation (18). The simplest is to say that when $\omega_D > \omega_c$ the electrons diffuse out of the positive trough of a wave in a time short compared with ω_c^{-1}, the dielectric relaxation time. It is as if the time allowed for relaxing the electric field of the sound wave were $(\omega_c/\omega_D)\omega_c^{-1}$ rather than ω_c^{-1}. Hence, only a fraction ω_c/ω_D of the maximum charge will be bunched in the troughs of the wave. In order to account for the diffusion factor ω_D/ω_c entering Equation (18) as a squared rather than a first-power term, we need to recognize that the work done by the drifting stream of electrons on the sound wave is proportional not only to the amount of charge bunched in the troughs but also to the asymmetry of the bunching. That is, to be completely effective, the bunched charge should be on the forward flank of the wave. Diffusion attenuates the magnitude of bunched charge, as cited above, by the factor ω_c/ω_D and, at the same time, attenuates the asymmetry by the same factor. Both effects are due to the tendency for diffusion to smear the bunched carriers into a uniform distribution. Hence, the effect on the acoustic gain enters in as $(\omega_c/\omega_D)^2$ just as the term $(\omega_r/\omega_c)^2$ in the denominator of Equation (18). The factor ω_r/ω_c in the numerator of Equation (18) reflects the increased rate of doing work as ω_r (i.e., v) increases. This effect is physically distinct from that due to the term $(\omega_r/\omega_c)^2$ in the denominator, which reflects the incomplete bunching.

Other ways of interpreting ω_D/ω_c come from writing this ratio in the following forms:

$$\frac{\omega_D}{\omega_c} = \frac{4\pi^2 kT\mu}{e\lambda^2} \times \frac{K}{4\pi ne\mu}$$

$$= \frac{\pi kTK}{ne^2\lambda^2}$$

$$= \left(\frac{2\pi kT\Delta n}{e\lambda n}\right)\left(\frac{1}{\mathcal{E}}\right), \quad \text{where} \quad \mathcal{E} = \frac{2\Delta ne\lambda}{K} \approx \text{electric field of sound wave,} \tag{20}$$

$$= \left(\frac{kT\Delta n}{en}\right)\frac{1}{\Delta V}, \quad \text{where} \quad \Delta V = \frac{\Delta ne\lambda^2}{\pi K} \approx \text{potential trough of sound wave,} \tag{21}$$

$$= \left(\frac{2\pi\lambda_D}{\lambda}\right)^2, \quad \text{where} \quad \lambda_D = \text{Debye length} = \left(\frac{kTK}{4\pi ne^2}\right)^{1/2}. \tag{22}$$

Equation (20) converts ω_D/ω_c into the ratio of effective diffusion field to the electric field of the sound wave. In Equation (21) this ratio is expressed as the ratio of the effective diffusion potential to the potential trough of the sound wave. In each case, the complete relaxation of the electric field of the sound wave is opposed by diffusion phenomena expressed either as a field or voltage.

In Equation (22) the ratio ω_D/ω_c is expressed as the square of the ratio of a Debye length to $\lambda/(2\pi)$, reflecting the well known fact that electric fields cannot be relaxed by conduction processes down to distances less than a Debye length.

3-D—Ratio of Electrical to Total Energy

(1) Meaning of Electrical and Total Energy

Consider for conceptual convenience a standing sound wave in a solid. At some phase in the oscillation of the sound wave all of the atoms will be at their maximum excursion and stationary. The energy of the sound wave is then entirely potential energy. At a later phase the atoms in the solid have zero displacement from their normal sites. The energy of the sound wave is then entirely the kinetic energy of moving atoms. We have called this kinetic energy the peak mechanical energy of the sound wave. The peak mechanical energy is clearly also the total energy of the sound wave.

At the phase when all of the atoms are at their maximum displacement, the total energy of the sound wave has been converted from kinetic energy to potential energy. We then compute how much of the potential energy can be dissipated by relaxation of the free carriers to lower energy states. This energy we have called the peak electrical energy and its ratio to the total energy (that is, the peak kinetic energy) we have called β, the ratio of electrical to total energy.

In this way we can distinguish between the general meaning of "electrical" and our restricted usage of it. The potential energy of a sound wave, for example, is finally an "electrical" energy since it is derived from the forces between neighboring ions and their valence electrons. But only a fraction of this energy is available for dissipation by the relaxation of free carriers, and it is this fraction that our restricted use of electrical energy refers to.

According to Equations (15) and (18), the distinction between the acoustoelectric gain constants for the several types of sound waves is contained in the coefficient β. The factor β enters in because the energy exchange between the sound wave and the free electrons is effected through the electrical energy accompanying the sound wave. At most, only the *electrical* energy of the wave can be doubled in one period of the wave. However, the energy needed to double the amplitude of sound wave is measured by the *total* energy of the wave. Maximum gain is achieved when $\beta \rightarrow 1$, that is, when the electrical energy approaches equality with the total energy. As we shall see, this tends to occur both for the polar optical modes of highly polar solids and for the high-frequency limit of deformation-potential sound waves.*

(2) Polar Optical Waves

For ionic crystals, it has been shown[10] that the longitudinal and transverse modes of vibration are connected by the relation

$$\frac{\omega_t^2}{\omega_l^2} = \frac{\epsilon_\infty}{\epsilon_0}. \tag{23}$$

The energy of the transverse modes is almost purely elastic energy, since the average electric field of the transverse modes is substantially zero. By Maxwell's equations, the transverse field (which arises from the time variation of currents of ions) is negligible owing to factors of order v_s/c, where v_s and c are the velocities of sound and light, respectively. The potential energy of the *longitudinal modes,* however, is the sum of an elastic energy comparable with that of the transverse modes *and* an electrical energy arising from the relative displacement of positive and negative ions. Hence, we can rewrite Equation (23) in the form

$$\beta = \frac{\text{peak electrical energy}}{\text{total energy}} = \frac{\omega_l^2 - \omega_t^2}{\omega_l^2} = \frac{\epsilon_0 - \epsilon_\infty}{\epsilon_0}. \tag{24}$$

Note that for a crystal with point ions ($\epsilon_\infty = 1$), $\beta \doteq 1$ as one would expect. Here the total energy when the ions are at their maximum displacement is electrostatic potential energy. Subsequently, when the ions are at their normal sites, the total energy is the kinetic energy of the ions and is equal to the electrical potential energy that the ions had at their maximum displacement.

We note further that as the crystal changes from a purely ionic solid, $\epsilon_\infty \rightarrow \epsilon_0$ and the ratio of electrical to total energy as measured by Equation (15) approaches zero. For such nonpolar solids a new source of electrical energy enters in, as in germanium or silicon, to be derived from the deformation potential of the optical modes.[11]

* This expression denotes sound waves coupled to free carriers via a deformation potential.

[10] R. H. Lyddane, R. G. Sachs, and E. Teller, "On the Polar Vibrations of Alkali Halides," *Phys. Rev.*, Vol. 59, p. 673, April 15, 1941.

[11] R. H. Parmenter, "Uniform Strains and Deformation Potentials," *Phys. Rev.*, Vol. 99, p. 1767, Sept. 15, 1955.

When the value of β in Equation (23) is inserted in Equation (18), Gunn's expression[6] for the acoustoelectric effect for polar optical waves is essentially reproduced. The reproduction would be exact if Gunn had used the common approximation of a single frequency for his optical phonons.

(3) Piezoelectric Acoustic Waves

The ratio of electrical to total energy of a piezoelectric acoustic wave is given approximately by the square of the electromechanical coupling constant;

$$\beta_{PE} = \frac{\epsilon_p{}^2}{KC}, \tag{25}$$

where ϵ_p = piezoelectric constant, C = elastic modulus, and K = relative dielectric constant. In CdS, for example, $\beta_{PE} \approx 0.05$. Other materials that have appreciable values of β_{PE} are ZnO, GaAs, and quartz.

When the value of β_{PE} given by Equation (25) is inserted in Equation (18), the result of Hutson and White is reproduced.

While Equation (25) is a well-known expression, a short derivation of it is inserted here for convenience. We begin with the piezoelectric relations

$$T = CS - \epsilon_p \mathcal{E}, \tag{26}$$

$$D = K\mathcal{E} + \epsilon_p S, \tag{27}$$

where T is the applied stress, S the strain, and D the electric flux. We set $D = 0$, solve for $\mathcal{E}$ in Equation (27), and substitute the result in Equation (26) to get

$$T = CS + \frac{\epsilon_p{}^2}{K} S. \tag{28}$$

If we multiply both sides by $S/2$,

$$\frac{1}{2} ST = \frac{1}{2} CS^2 + \frac{1}{2}\frac{\epsilon_p{}^2}{K} S^2; \tag{29}$$

the first term on the right is an elastic energy and the second the electrical energy associated with the piezoelectric effect. The ratio of the electrical to the elastic energy terms reproduces Equation (25).

Since the electrical energy is small, this ratio is a good approximation to the ratio of electrical to total energy.

(4) Deformation-Potential Acoustic Waves

In order to derive a value for the electrical energy to be ascribed to deformation-potential acoustic waves, we make use of the equivalent electrical field $\mathcal{E}$ obtained from the spatial derivative of the potential due to deformation. One may question whether this is a real field in the sense that it should obey Maxwell's equations. However, there is no question that an electron moves in this field as if it were real. Hence, the electrical energy we are computing is the energy that is obtained when electrons slide into a potential trough in sufficient numbers to cancel the applied equivalent field. In brief, the electrical energy is $K\mathcal{E}^2/(8\pi)$ as if $\mathcal{E}$ were a real field.

The potential due to a deformation S is

$$V_D = \frac{B}{e} S, \tag{30}$$

where B is called the deformation potential,* e the charge on an electron, and S the mechanical strain. For an acoustic wave,

$$S = S_0 \sin\left(\frac{2\pi}{\lambda} x + \omega t\right). \tag{31}$$

The maximum field due to the deformation caused by the acoustic wave is, from Equations (30) and (31):

$$\mathcal{E}_0 = \left.\frac{dV_D}{dx}\right|_{\max} = \frac{2\pi B}{e\lambda} S_0. \tag{32}$$

The maximum strain, S_0, can be written as

$$S_0 = \frac{2\pi}{\lambda} A, \tag{33}$$

* Note that B has the dimensions of energy (ergs) even though by convention it is called a "potential."[12]

[12] W. Shockley and J. Bardeen, "Energy Bands and Mobilities in Monatomic Semiconductors," *Phys. Rev.*, Vol. 77, p. 407, Feb. 1, 1950.

where A is the mechanical displacement amplitude of the wave. Hence Equation (32) becomes

$$\mathcal{E}_0 = \frac{4\pi^2 B}{e\lambda^2} A. \tag{34}$$

The peak electrical energy per unit volume is then

$$\begin{aligned} E_{elect} &= \frac{K\mathcal{E}_0^2}{8\pi} = \frac{KB^2}{8\pi e^2}\left(\frac{4\pi^2}{\lambda^2}A\right)^2 \\ &= \frac{KB^2\omega^4A^2}{8\pi e^2 v_s^4}. \end{aligned} \tag{35}$$

The total energy is given by the peak kinetic energy per unit volume;

$$E_{tot} = \frac{1}{2}\rho A^2\omega^2, \tag{36}$$

where ρ is the density of the medium. From Equations (35) and (36) we get

$$\frac{E_{elect}}{E_{tot}} = \beta_{AD} = \frac{KB^2\omega^2}{4\pi\rho e^2 v_s^4}. \tag{37}$$

For long-wavelength phonons $\beta_{AD} \to 0$ since $\omega^2 \to 0$. At the other end of the spectrum, namely for $\omega = 10^{14}$ rad/sec, β_{AD} becomes of the order of unity for values of B of the order of ten electron volts per unit strain. [In Equation (32) the units of B are ergs per unit strain; the electronic charge is 4.7×10^{-10} esu.] This estimate is significant because it raises the question of whether β can, in principle, exceed unity. We will return to this question in Section 3-E(4) on spontaneous deformation.

(5) Nonpolar Optical Waves

Optical waves in nonpolar materials such as germanium and silicon do not have the obvious electric fields associated with them that the optical waves in ionic solids have. For our purposes an effective electric field can be constructed out of the concept of the optical deformation potential, first introduced by Parmenter,[11] which was defined as the energy shift of the band edge per unit displacement of the two sublattices relative to each other. Conwell[13] quotes a value for the conduction band of germanium of 4×10^8 electron volts per centimeter.

We wish to compute the ratio of electrical to total energy (β_{OD}) for the nonpolar optical modes. If we take D to be the optical deformation potential, we can write, for the maximum effective electric field of a wave,

$$\mathcal{E}_0 = \frac{2\pi DA}{e\lambda}, \tag{38}$$

where A is the amplitude of displacement of the two sublattices relative to each other.

The peak energy density to be associated with the electric field is then

$$E_{elect} = \frac{K\mathcal{E}_0^2}{8\pi} = \frac{\pi K D^2 A^2}{2e^2\lambda^2}. \tag{39}$$

Here it is understood that the electrical energy is derived from the flow of electrons into the potential troughs.

As before, the total energy is given by energy density of the peak kinetic energy;

$$E_{tot} = \frac{1}{2}\rho A^2\omega^2. \tag{40}$$

Then

$$\frac{E_{elect}}{E_{tot}} = \beta_{OD} = \frac{\pi K D^2}{\rho e^2\omega^2\lambda^2}. \tag{41}$$

Parenthetically, it is worth noting that $\beta_{OD} \approx 1$ for $K = \rho = 10$, $D = 4 \times 10^8$ ev/cm, $\omega = 5 \times 10^{13}$ rad/sec, and $\lambda = 6 \times 10^{-8}$ cm.

(6) Intervalley Deformation-Potential Waves

The ratios of electrical to total energy that have been computed thus far for the several types of sound waves have all been limited by the amount of charge that can be accumulated in a potential trough of the sound wave. This charge was limited by the actual or effective electric field in the wave. In contrast to this spatial type of bunching of electrons, Holstein, as quoted by Weinreich,[14] has proposed that for

[13] E. M. Conwell, "Relative Energy Loss to Optical and Acoustic Modes of Electrons in Avalanche Breakdown in Ge," *Phys. Rev.*, Vol. 135, p. A1138, 17 Aug. 1964.

[14] G. Weinreich, T. M. Sanders, Jr., and H. G. White, "Acoustoelectric Effect in n-Type Germanium," *Phys. Rev.*, Vol. 114, p. 33, April 1, 1959.

deformations that remove the degeneracy of equivalent valleys in the energy band structure, a bunching in phase space will take place. That is, electrons that populate equivalent valleys equally in unstrained material will tend to populate the lower of the two valleys (Figure 4) when the material is strained and the valleys are energetically separated. The energy to be gained from the repopulation of valleys is proportional to the density of electrons and is not constrained by the

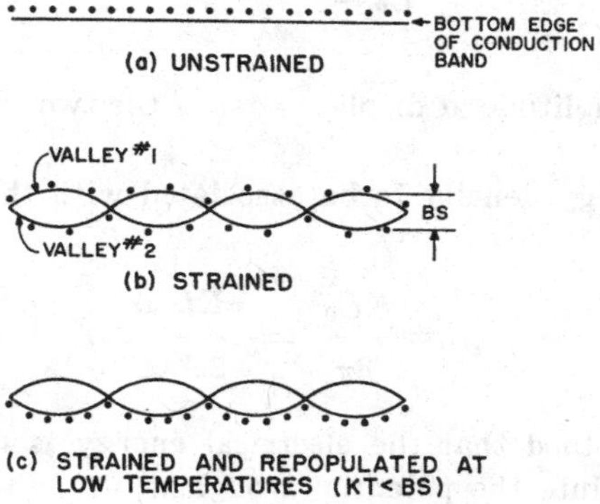

Fig. 4—Schematic diagram of electrical energy gained from a relative shift of valleys due to a shear deformation.

space charge that would arise from a spatial bunching. Pomerantz[15] gives the following expression for ω_a (translated into our terminology):

$$\omega_a = \left[\frac{nB^2}{9\rho v_s{}^2 kT}\right] \omega \frac{\omega_r/\omega_c}{1+(\omega_r/\omega_c)^2}, \tag{42}$$

where n is the density of free electrons, B the deformation potential (energy spread between valleys per unit strain), and $\omega_c = \tau_c{}^{-1}$, where τ_c is the relaxation time for repopulation of the valleys.

Equation (42) matches our general relation, Equation (15), if we can show that the bracketed part of Equation (42) is equal to β, the ratio of electrical to total energy. To do so we write the electrical energy per unit volume as

$$E_{elect} = \frac{nBS}{2}\left(\frac{BS}{kT}\right). \tag{43}$$

[15] M. Pomerantz, "Amplification of Microwave Phonons in Germanium," *Phys. Rev. Letters*, Vol. 13, p. 308, 31 Aug. 1964.

Here $nBS/2$ is the electrical energy that would be gained if all of the electrons in the upper valley dropped into the lower valley. Since, however, the deformation energy BS is small compared with kT, only the fraction $BS/(kT)$ will undergo repopulation. For degenerate material, the fraction is $(3/2)(BS/E_f)$, where E_f is the Fermi energy.

The total energy, as before, is

$$E_{tot} = \frac{1}{2}\rho\omega^2 A^2 = \frac{1}{2}\rho v_s^2 \left(\frac{2\pi A}{\lambda}\right)^2 = \frac{1}{2}\rho v_s^2 S^2. \tag{44}$$

From Equations (43) and (44),

$$\beta_{VD} = \frac{E_{elect}}{E_{tot}} = \frac{nB^2}{\rho v_s^2 kT}. \tag{45}$$

This matches the bracketed part of Pomerantz's Equation (42) except for the numerical factor of 9. The numerical factor arises because our value of B is defined for valleys lying in the [100] direction while the valleys in germanium lie in the [111] direction. When B is resolved along the [111] direction the resultant deformation constant is $B/3$ (see Reference (14)).

(7) Metals

In the case of metals, the major source of electrical energy comes from the change in kinetic energy of the electrons when the lattice is compressed or expanded by a sound wave. Consider, for example, the compression half of a longitudinal sound wave (Figure 5b). The first effect of the compression is to increase the kinetic energy of the component of electron velocity along the line of propagation. The increase is a direct result of the increased density of electrons. After an energy relaxation time τ_c, determined by collisions with the lattice, the excess energy along the direction of propagation is distributed among all three dimensions, so that there is a common Fermi level (Figure 5c) somewhat higher than the thermal equilibrium Fermi level. It is this relaxation energy that constitutes the electrical energy of the system available for dissipation of the total energy of the sound wave (see Section 3-D(1)).

The magnitude of the electrical energy can be estimated as follows. Let an element of volume of the metal suffer a strain S along each direction so that the density of electrons will be

$$n \doteq n_0 (1 + 3S), \tag{46}$$

and the Fermi level will be

$$E_f = An^{2/3} = An_0^{2/3}\,(1+3S)^{2/3}$$

$$\doteq E_{f0}\,(1+2S). \qquad (47)$$

Here n_0 and E_{f0} are the unstrained thermal equilibrium values.

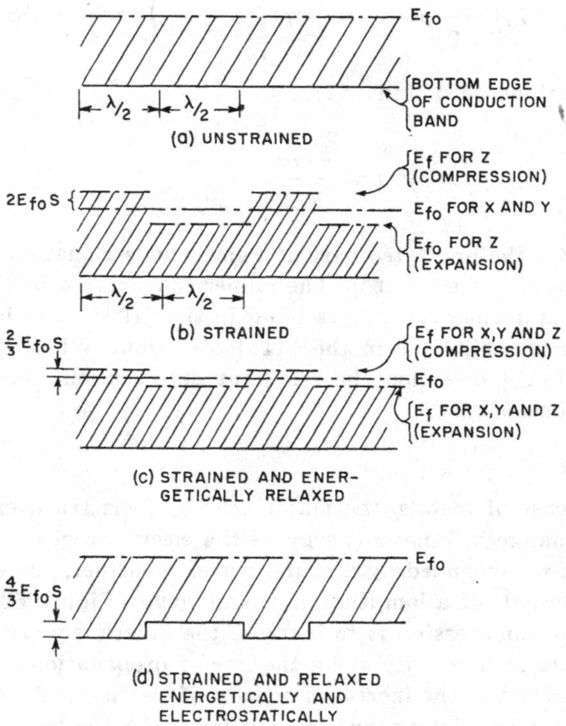

Fig. 5—Schematic diagram of electrical energy gained when a metal is deformed by a periodic wave.

Now if we remove the strain from two of the axes and retain it on the third, the above relations will remain valid for the third axis. Hence we have an excess density of kinetic energy along one axis. This energy will be relaxed until it is distributed along all three axes. The excess energy before relaxation is, from Equation (47),

$$\frac{3}{2}\left[\left(\frac{1}{3}\right)\left(\frac{1}{2}\right)\frac{(2E_{f0}S)}{E_{f0}}\,n_0\right] \times [2E_{f0}S] = n_0E_{f0}S^2. \qquad (48)$$

The first square bracket is the number of electrons lying in the interval between E_{f0} and E_f, that is, in the energy range $2E_{f0}S$.* The factor 1/3 is due to counting electrons along one axis only. The factor 1/2 accounts for averaging the energy over the excess electrons. The second square bracket is the height by which the Fermi level was raised.

The excess energy after relaxation to a common Fermi level (see Figure 5c) is computed as in Equation (48) using all of the electrons and an energy, $E_f - E_{f0} = (2/3)E_{f0}S$. The result is 1/3 the result for Equation (48). Hence, the electrical energy is

$$E_{elect} = \left(1 - \frac{1}{3}\right) n_0 E_{f0} S^2 = \frac{2}{3} n_0 E_{f0} S^2, \tag{49}$$

and the ratio of electrical to total energy is

$$\beta_M = \frac{\frac{2}{3} n_0 E_{f0} S^2}{\frac{1}{2} \rho v_s^2 S^2} = \frac{4}{3} \frac{n_0 E_{f0}}{\rho v_s^2}. \tag{50}$$

In a simplified model of a metal, as shown by Bohm and Staver[16] and quoted by Pines,[17] the following relation holds:

$$\frac{1}{2} \rho v_s^2 = \frac{1}{3} n_0 E_{f0}. \tag{51}$$

From Equations (51) and (50),

$$\beta_M = 2. \tag{52}$$

This estimate of β_M is probably high, owing to the approximate nature of Equation (51). If true, it would lead to a gross instability of the metal (see Section 3-E(4)). The approximation, however, is sufficient for our immediate purpose.

---

* Since $n \propto E_f^{3/2}$ it follows that $\Delta n/n = (3/2)\Delta E_f/E_f$.

16 D. Bohm and T. Staver, "Application of Collective Treatment of Electron and Ion Vibrations to Theories of Conductivity and Superconductivity," *Phys. Rev.*, Vol. 84, p. 836, Nov. 15, 1951.

17 D. Pines, "Electron Interaction in Metals," *Solid State Physics*, Edited by F. Seitz and D. Turnbull, Vol. 1, p. 367, Academic Press, New York, 1955.

Equation (52) is inserted into Equation (15) to obtain

$$\omega_a = 2\omega \frac{\omega_r/\omega_c}{1 + (\omega_r/\omega_c)^2}. \tag{53}$$

While Equation (53) shows that, in principle, sound waves can be amplified by a drifting stream of carriers in a metal, such amplification is not likely, in practice, since drift velocities in excess of 3×10^5 cm/sec, corresponding to current densities in excess of about 10^9 amperes/cm^2, are needed. A more realistic test of Equation (53) is to evaluate it for the attenuation of sound waves, namely for $|\omega_r| = \omega$. Also we take $\omega_c = \tau_c^{-1} \gg \omega$. Equation (53) then becomes

$$\omega_a = 2\omega^2\tau_c. \tag{54}$$

Equation (54) is to be compared with the result obtained by Pippard[18] or, more recently, by Levy[19];

$$\omega_a = \frac{4}{15} \frac{mn}{\rho\tau_c} (ql)^2, \tag{55}$$

where $q = 2\pi/\lambda$, $l = v_0\tau_c$ and $(1/2)\ mv_0^2 = E_f$. These values substituted into Equation (55) yield

$$\omega_a = \frac{8}{15} \frac{nE_f\omega^2\tau_c}{\rho v_s^2}. \tag{56}$$

Pippard points out that τ_c is the velocity relaxation time and is twice as large as the energy relaxation time, which we have used. Hence, for comparison, Equation (56) should be multiplied by a factor of 2 to convert τ_c into an energy relaxation time.

Again making use of Equation (51), Equation (56) becomes

$$\omega_a = \frac{8}{5} \omega^2\tau_c. \tag{57}$$

Our Equation (49) is in close agreement with Equation (57).

---

[18] A. B. Pippard, "Ultrasonic Attenuation in Metals," *Phil. Mag.*, Vol. 46, p. 1104, Oct. 1955.

[19] M. Levy, "Ultrasonic Attenuation in Superconductors for ql < 1." *Phys. Rev.*, Vol. 131, p. 1497, 15 Aug. 1963.

Equation (57) is for longitudinal sound waves. The result for transverse sound waves, according to Pippard and Levy, is the same except that the numerical factor is 10% smaller. The substantial equivalence of longitudinal and transverse waves is not unreasonable, since the pattern of raising and lowering of Fermi levels that holds for successive half wavelengths of a longitudinal wave is reproduced in each half wavelength of the transverse wave. A shear distortion results in the increase in one dimension and a decrease in an orthogonal dimension without a change in volume.

Note that in the case of longitudinal waves the transfer of charge between the two half wavelengths needed to bring their Fermi levels to the same value (Figure 5d) contributes a negligible energy at long wavelengths but can be significant at wavelengths approaching lattice dimensions.

3-E—General Remarks

(1) Comparative Effects

It is clear that the parameter electrical energy/total energy assesses directly the relative strengths of the interaction between electrons and the several types of sound waves. At long wavelengths ($\lambda \geqq 10^{-5}$ cm), for example, the interactions with acoustic and nonpolar optical deformation-potential waves rapidly become small compared with piezoelectric or polar optical waves, since the ratios of electrical to total energy of the former decrease as λ^{-2}. The electrical energy decreases at long wavelengths owing to the reduction in charge that can be accumulated in the potential troughs. The reduction in charge arises from the reduced electric fields at long wavelengths. For piezoelectric and polar optical waves, the ratio of electrical to total energy is a constant independent of wavelength since the electric field is proportional to the elastic strain. In the case of the electrical energies arising from the relative shift of equivalent valleys, there is no spatial bunching of electrons, so that the ratio of electrical to total energy is again a constant independent of wavelength.

Figure 6 shows some representative curves for the ratio of electrical to total energy versus the reciprocal of the wavelength of the various types of sound waves. The curves for deformation-potential acoustic and optical waves exceed unity at the shortest wavelengths. It is questionable, of course, whether the deformation potential retains its validity in this range. At these short wavelengths the densities of carriers needed to relax the electric fields lie in the range of degeneracy. For effective relaxation the Debye length should be less than $\lambda/2$, that is,

$$\left(\frac{KkT}{4\pi ne^2}\right)^{1/2} \leqq \frac{\lambda}{2},$$

or

$$n \geqq \frac{KkT}{\pi e^2 \lambda^2}. \tag{58}$$

This leads to degeneracy at room temperature for $\lambda \leqq 10^{-6}$ cm.

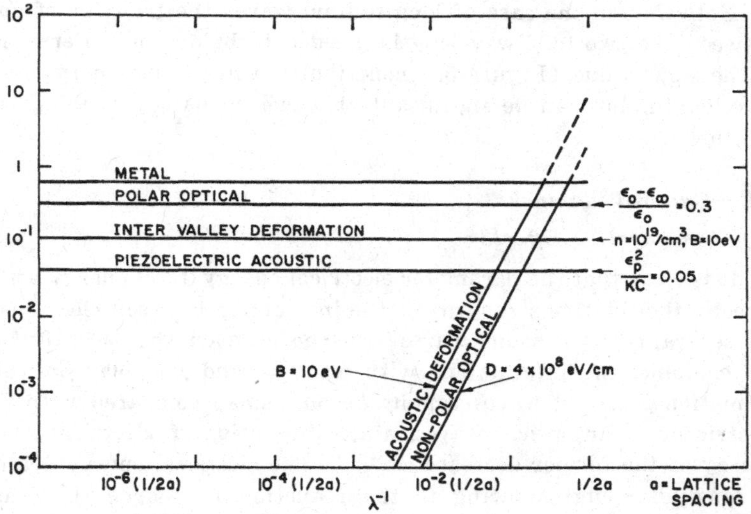

Fig. 6—Plot of β, the ratio of electrical to total energy, as a function of reciprocal wavelength for the various types of phonons. Parameters used in calculating these curves other than those shown on the figure are: $K = 10$, $\omega_{max} = 5 \times 10^{13}$ rad/sec, $\rho = 5$ grams/cm^3, $v_s = 5 \times 10^5$ cm/sec, $T \doteq 300°$K.

The interaction of electrons with polar optical waves appears to be one of the strongest, since their ratio of electrical to total energy may be as high as 0.5 compared, for example, with 0.05 for piezoelectric waves or 0.1 for the intervalley deformation potential waves. The difficulty, however, is that the drift velocity of the electrons must exceed the *phase* velocity of the optical phonons. The latter increases toward long wavelengths as $\lambda\omega/(2\pi)$ where $\omega \approx 10^{14}$ rad/sec. Hence, for $\lambda \approx 10^{-6}$ cm, the phase velocity already exceeds 10^7 cm/sec—a reasonable upper limit to the drift velocity in most materials. A further limitation to the interaction with optical waves in accordance with Equation (18) is that the validity of this equation was based on the implied assumption that the mean free path of carriers was less than

$\lambda/(2\pi)$. Longer mean free paths would lead to a quantum rather than classical approach. If we take $\lambda \approx 10^{-6}$ cm as the longest reasonable wavelength with which carriers can interact (based on a drift velocity of 10^7 cm/sec), their mobility mean free path must be less than 2×10^{-7} cm and their mobility accordingly must be less than about 20 cm^2/volt-sec. The result is that Equation (18) has a rather narrow region of validity for optical waves. It lies in the neighborhood of $\lambda \approx 10^{-6}$ cm, carrier drift velocity $\approx 10^7$ cm/sec, and mobility ≈ 10 cm^2/volt-sec.

(2) Transverse Electric Fields

The method used here for computing the acoustoelectric gain (or loss) depends upon the presence of a relaxable field or energy associated with the sound wave. Hence, the method is valid for longitudinal electric fields and for the relaxation of energy involved in the repopulation of shifted valleys. It is not immediately evident that the method is also valid for electric fields transverse to the direction of propagation. Such fields are in any event almost vanishingly small for sound waves since, by Maxwell's equations, they must arise from a time rate of change of magnetic field associated with moving ions. Hence, these transverse fields are in the order of (velocity of sound)/(velocity of light) smaller than the longitudinal fields arising from the ionic charges themselves.

We do know that the transverse component of electric field of a sound wave can cause an attenuation of the energy of the sound wave at the rate $\mathcal{E}^2_{trans}\sigma$. Moreover, since this rate of dissipation is a function of the relative velocities of the sound wave and the drifting carriers we can still make use of Equations (5)-(6) to derive an acoustoelectric effect. For example, an electron drifting with the phase velocity of a sound wave should see a "static" surround, that is, no time-varying currents or magnetic fields and hence no electric field. For such drift velocities the dissipation should vanish. Similarly, electrons drifting much faster than the phase velocity of the sound wave should see a higher transverse electric field than that seen by nondrifting electrons and should lead to a higher rate of dissipation and, by Equations (5)-(6), to amplification of the sound wave.

Note that the transverse electric fields of a sound wave can also be relaxed by the motion of electrons but not in the same way longitudinal fields are relaxed. The transverse fields are derived from the time-varying magnetic fields of the ionic currents. If the electrons keep pace with the ions, the net current, and hence the transverse electric field, will approach zero. This is the situation that obtains in metals.

A more pertinent example of pure transverse fields is that of electromagnetic radiation. The absorption of light by free carriers leads to a d-c current since, by the well-known phenomenon of radiation pressure, the momentum of the light wave is first absorbed by the free carriers and then passed on to the crystal lattice. If, however, the electrons drift with the speed of light in the medium, they should see zero electric field since, by Maxwell's equations, the force on an electron is

$$F = e\left[\mathcal{E} - \frac{v}{c}H\right]$$

$$= e\left[\mathcal{E} - \frac{c}{\epsilon_\infty{}^{1/2}c}\epsilon_\infty{}^{1/2}\mathcal{E}\right] = 0, \qquad (59)$$

since $H^2 = \epsilon_\infty \mathcal{E}^2$.

By the same argument, electrons drifting faster than the speed of light in the medium should see an increasing transverse electric field and suffer increasing dissipation and hence, by Equations (5)-(7), should amplify a coherent beam of light or a radio wave. (See Appendix II by Lampert.)

(3) Acoustoelectric Currents

Our discussion has emphasized the amplification of sound waves by a drifting stream of carriers. On the other hand, when the carriers are stationary the same expression (Equation (18)) gives the attenuation or rate of loss of energy of the acoustic waves. By Equation (6) we can convert the rate of loss of energy of the sound wave into an equivalent electric field acting on the free carriers;

$$\omega_a E_{tot} = F v_s = \mathcal{E} e n v_s. \qquad (60)$$

Here E_{tot} is the energy density of the sound wave, n the density of the free carriers, and $\mathcal{E}$ the equivalent electric field acting on the carriers. The acoustoelectric current would then be $\mathcal{E} n e \mu$. It was this current and field that Parmenter identified in his pioneer paper as the acoustoelectric current and field.

(4) Spontaneous Deformation

If the electrical energy to be gained from the relaxation of electrons in a deformed lattice exceeds the work required to effect the deformation, the lattice will spontaneously deform to a lower energy state. The criterion for this spontaneous deformation is, by definition, $\beta > 1$.

Such an inequality is logically impossible for polar optical waves and piezoelectric waves because the form of β is

$$\beta = \frac{\text{electrical energy}}{\text{total energy}} = \frac{\text{electrical energy}}{\text{elastic energy} + \text{electrical energy}}.$$

That is, the energy of the electric field is itself a part of the total energy. In these cases there is a macroscopic, easily identifiable electric field.

In the case of the acoustic and the optical deformation-potential waves, however, it is not clear that the "field" that causes electrons to accumulate in the troughs of the waves is of such a character that its energy should be added to a separate elastic energy in order to obtain the total energy. If this were so, then, of course β would always be less than or equal to unity. If this were not so, then β could possibly exceed unity and lead to a spontaneous deformation of the lattice. Since β approaches unity only at the shortest wavelengths, the spontaneous deformation would occur only for metallic densities of electrons. The short-wavelength deformation of the lattice would then give rise to a small forbidden gap at the Fermi level. This is the type of model that Frohlich[20] proposed in 1953 for a superconductor.

Even when the electrical energy is less than the total energy, its presence can still be observed as a change in the elastic constant and, hence, of the velocity of sound. The electrical energy makes it easier to deform the solid. Such an effect was reported by Bruner and Keyes[21] for degenerate germanium with a carrier concentration of 3×10^{19}/cm^3. The effect was a 5 per cent decrease in the elastic constant. The electrical energy was derived from the relative energy shift of equivalent valleys. The electrical energy for intervalley deformation is given by Equation (45) for nondegenerate materials. Degeneracy is taken into account by replacing the factor $BS/(kT)$ by $(3/2)(BS/E_f)$. The resulting expression is

$$\beta_{VD} = \frac{3n_0B^2}{2\rho v_s{}^2E_f}, \tag{61}$$

[20] H. Fröhlich, "On the Theory of Superconductivity: the One-Dimensional Case," *Proc. Roy. Soc.* (London), Vol. A223, p. 296, 6 May, 1954.

[21] L. J. Bruner and R. W. Keyes, "Electronic Effect in the Elastic Constants of Germanium," *Phys. Rev. Letters*, Vol. 7, p. 55, July 15, 1961.

which matches Bruner and Keyes' expression within a numerical factor of order unity.

(5) Recapitulation

Table I lists the time constant ω_a for amplification or attenuation of the various sound waves. Several comments on Table I need to be made.

(a) ω_a is the time rate of change of energy for the various sound waves, normalized to their energy density. For example, ω_a^{-1} is the time required to approximately double the energy density of the sound wave in the case of amplification ($\omega_r > 0$) or to halve the energy density in the case of attenuation ($\omega_r < 0$).
(b) ω_a can be converted into a space rate of change of energy by dividing by the group velocity v_g.
(c) The maximum value of ω_a is approximately ω, the frequency of the sound wave, and occurs for $\beta = 1$ and $\omega_r = \omega_c$.
(d) All of the expressions are taken from the published literature cited with the exceptions of the deformation potential acoustic and the nonpolar optical waves. For the acoustic waves there was no convenient expression with which to compare our results. For the optical waves, no published discussion was found.
(e) The dielectric constants for the polar optical waves are specifically labeled as ϵ_0 for the low-frequency value and ϵ_∞ for the high-frequency or optical value. In the case of the acoustic waves the symbol K is used for the dielectric constant. Its value will be a function of the doppler shifted frequency, ω_r. However, for non-polar materials the dielectric constant varies by considerably less than a factor of two from low to high frequencies.
(f) The velocity of sound v_s is the phase velocity and for optical waves this velocity rapidly takes on large values as the wavelength increases. For optical waves $v_s \approx 10^{13}\lambda$ cm/sec.
(g) The arguments used to obtain most of the expressions in Table I involve the conductivity, σ, of the medium. In order for the conductivity to retain its d-c value, the collision time of the electron should be less than ω_r^{-1}. This condition is not appropriate for the case of intervalley deformation and metals where the collision time is itself the energy relaxation time.
(h) The diffusion term does not appear in the cases of intervalley deformation and of metals, because the energy relaxation processes considered do not involve a spatial redistribution of electrons.

Table II lists the time rates of energy loss by fast electrons to the various types of phonons and to plasmons and Cerenkov radiation. The physical arguments for deriving these expressions will be given in Part II of this paper. These expressions have been taken out of the literature and recast algebraically in a common form to show that the key parameter for the energy losses to phonons is the ratio of electrical to total energy for the various phonons. It is this same parameter that characterizes the several acoustoelectric effects listed in Table I. In the case of energy losses to electronic excitations, β satisfies the more general definition, as will be discussed in Part II, namely, the fraction of available coulomb energy used in forming an energy well.

Appendix I*—Interchange of Energy between a Longitudinal Electric Wave and Drifting Carriers in a Solid

Consider a longitudinal electric wave in a solid in the absence of conduction electrons (or holes):

$$\vec{\mathcal{E}} = \hat{z}\mathcal{E}_0 \exp i(\omega t - kz). \tag{62}$$

For example, $\vec{\mathcal{E}}$ could be the electric field associated with a piezo-active acoustic wave in CdS.

In the presence of conduction electrons the "impressed" field $\mathcal{E}_0$ will be partially relaxed to a new value $\mathcal{E}_e$, which can be written

$$\mathcal{E}_e = \mathcal{E}_0 - \mathcal{E}_i, \tag{63}$$

where $\mathcal{E}_i$ is the field induced by the relaxation, i.e., by space charge ρ_i created in the relaxation process.

If further the conduction electrons are drifted in the direction $+z$ of the wave at velocity v, then the angular frequency "seen" by the electrons is

$$\omega_r = \omega \frac{|v - v_s|}{v_s}, \qquad v_s = \frac{\omega}{k}, \tag{64}$$

and all a-c quantities, in the frame of reference moving with the electrons, have the space–time variation $\exp i(\omega_r t - kz)$. Hence the operators $\nabla\cdot$ and $\partial/\partial t$, in this frame of reference, become $-ik\hat{z}\cdot$ and $i\omega_r$, respectively.

* The Appendixes to this paper were written by M. A. Lampert.

Table I—Time Constant (ω_a) for Rate of Change of Energy Density in Sound Waves

Phenomenon	ω_a	Source
Polar Optical	$\left[\dfrac{\epsilon_0 - \epsilon_\infty}{\epsilon_0}\right] \omega f(\omega_r, \omega_c, \omega_D)$	Gunn[6] Woodruff[7]
Piezoelectric Acoustic	$\left[\dfrac{\epsilon_p{}^2}{KC}\right] \omega f(\omega_r, \omega_c, \omega_D)$	Hutson, McFee and White[5]
Deformation Potential Acoustic	$\left[\dfrac{KB^2\omega^2}{4\pi\rho e^2 v_s{}^4}\right] \omega f(\omega_r, \omega_c, \omega_D)$	Parmenter[2] Weinreich[3]
Nonpolar Optical	$\left[\dfrac{\pi K D^2}{\rho e^2 \omega^2 \lambda^2}\right] \omega f(\omega_r, \omega_c, \omega_D)$	
Intervalley Deformation Potential	$\left[\dfrac{nB^2}{\rho v_s{}^2 kT}\right] \omega f(\omega_r, \omega_c)$	Weinreich[3] Pomerantz[15]
Metals	$[\approx 1] \quad \omega f(\omega_r, \omega_c)$ or $\omega^2 \tau_c$ for $\omega_r = \omega$ and $\omega\tau_c \ll 1$	Pippard[18] Levy[19]

Note that the above expressions are valid for the mean free path of the electrons less than the wavelength of the sound wave.

Definitions for Table I

$$\omega_a = \frac{1}{E}\frac{dE}{dt}; \quad E = \text{energy density of sound wave}$$

$$f(\omega_r, \omega_c) = \frac{\omega_r/\omega_c}{1 + (\omega_r/\omega_c)^2},$$

$\omega_r = \omega(v_d - v_s)/v_s$,

$\omega_D = 4\pi^2 kT\mu/(e\lambda^2)$,

$\omega_c = 4\pi\sigma/K$, or $\omega_c = \tau_c{}^{-1}$ for metals and intervalley deformation, where τ_c is the time for an electron to come to thermal equilibrium,

v_d = drift velocity of electrons,

v_s = phase velocity of sound,

ϵ_0 = low-frequency dielectric constant,

ϵ_∞ = high-frequency (optical) dielectric constant,

K = dielectric constant,

Definitions for Table I (continued)

ϵ_p = piezoelectric constant,

C = elastic modulus,

B = deformation potential (electron volts in ergs/unit strain),

D = optical deformation potential (electron volts in ergs per centimeter relative shift of sublattices).

Table II—Time Rates of Loss of Energy by Electrons of Velocity v.

Phenomenon	dE/dt	Source
Polar Optical Phonons	$\left[\frac{\epsilon_0 - \epsilon_\infty}{\epsilon_0}\right]\frac{e^2\omega^2}{\epsilon_\infty v}\ln\left(\frac{2mv^2}{\hbar\omega}\right)$ Note: $\frac{1}{2}mv^2 > \hbar\omega$	Fröhlich[22] Callen[23]
Piezoelectric Phonons	$\frac{\pi}{4}\left[\frac{\epsilon_p^2}{KC}\right]\frac{e^2\omega^2}{Kv}$ Note: $2mv = \hbar\omega/v_s$	Tsu[24]
Acoustic Phonons	$\frac{1}{4}\left[\frac{B^2\omega^2K}{4\pi e^2\rho v_s^4}\right]\frac{e^2\omega^2}{Kv}$ Note: $2mv = \hbar\omega/v_s$	Seitz[25] Conwell[13]
Nonpolar Optical Phonons	$\frac{1}{2}\left[\frac{\pi KD^2}{\rho e^2\omega^2\lambda^2}\right]\frac{e^2\omega^2}{Kv}$ Note: $\frac{1}{2}mv^2 > \hbar\omega$ and $\frac{\lambda}{2\pi} = \frac{\hbar}{2mv}$	Conwell[13]
Electronic excitations	$\left[\frac{\omega_p^2}{\omega_e^2}\right]\frac{e^2\omega_e^2}{v}\ln\left(\frac{2mv^2}{\hbar\omega_e}\right)$ Note: $\omega_p = \frac{4\pi ne^2}{m} \leq \omega_e$ n = density of electrons whose excitation energy is $\hbar\omega_e$ $\frac{1}{2}mv^2 > \hbar\omega_e$	Bohr[26] Bethe[27]
Plasmons	$[1]\ \frac{e^2\omega^2}{v}\ln\left(\frac{2mv^2}{\hbar\omega}\right)$ Note: ω = plasma frequency and $\frac{1}{2}mv^2 > \hbar\omega$	Bohm and Pines[28]
Cerenkov	$\left[1 - \frac{1}{\epsilon_\infty}\right]\frac{e^2\omega^2}{v}$ Note: $v = c$ = velocity of light in vacuum	See, e.g., Schiff[29]

Definitions for Table II

ϵ_0 = low-frequency dielectric constant,

ϵ_∞ = high-frequency (optical) dielectric constant,

K = dielectric constant,

e_p = piezoelectric constant,

C = elastic modulus (dynes/cm^2),

ρ = density (grams/cm^3),

v_s = phase velocity of sound,

v = velocity of electron,

ω = angular frequency of sound wave,

B = deformation potential (electron volts in ergs/unit strain)

D = optical deformation potential (electron volts in ergs per centimeter relative shift of sublattices),

m = effective mass of electrons.

References for Table II

[22] H. Fröhlich, "On the Theory of Dielectric Breakdown in Solids," *Proc. Roy. Soc.* (London), Vol. A188, p. 521, 25 Feb. 1947.

[23] H. B. Callen, "Electric Breakdown in Ionic Crystals," *Phys. Rev.*, Vol. 76, p. 1394, Nov. 1, 1949.

[24] R. Tsu, "Phonon Radiation by Uniformly Moving Charged Particles in Piezoelectric Solids," *Jour. Appl. Phys.*, Vol. 35, p. 125, Jan. 1964.

[25] F. Seitz, "On the Theory of Electron Multiplication in Crystals," *Phys. Rev.*, Vol. 76, p. 1376, Nov. 1, 1949.

[26] N. Bohr, *Phil. Mag.*, Vol. 25, p. 10, 1913; Vol. 30, p. 581, 1915.

[27] H. Bethe, *Ann. Phys.*, Vol. 16, p. 285, 1933; *Handbuch der Physik* (Geiger and Scheel, eds.), Vol. 24, Part I, Springer, Berlin, 1933.

[28] D. Bohm and D. Pines, "A Collective Description of Electron Interactions: III. Coulomb Interactions in a Degenerate Electron Gas," *Phys. Rev.*, Vol. 92, p. 609, Nov. 1, 1953 (Quoted by L. Marton et al. in *Advances in Electronics*, L. Marton, ed., Vol. 7, p. 230, Academic Press, New York, 1955.)

[29] L. I. Schiff, *Quantum Mechanics*, p. 271, McGraw-Hill Book Co., New York, 1955.

The Poisson and continuity equations are, respectively, (MKS units)

$$\epsilon \nabla \cdot \vec{\mathcal{E}}_i = \rho_i \rightarrow \mathcal{E}_i = \frac{i}{k\epsilon} \rho_i \tag{65}$$

$$\nabla \cdot \vec{J} + \frac{\partial \rho_i}{\partial t} = 0 \rightarrow \rho_i = \frac{k}{\omega_r} J = \frac{k\sigma_0}{\omega_r} \mathcal{E}_e, \tag{66}$$

where we have used

$$\vec{J} = \sigma \vec{\mathcal{E}}_e, \quad \sigma = \frac{\sigma_0}{1 + i\omega_r\tau} \approx \sigma_0 = \frac{e^2 n_0 \tau}{m^*}, \qquad (\omega_r\tau \ll 1). \tag{67}$$

Note that the current flows in response to the *net* field, $\mathcal{E}_e$, acting on the charges.

Combining Equations (65), (66), and (63) we obtain

$$\mathcal{E}_i = i\frac{\omega_c}{\omega_r} \mathcal{E}_e, \qquad \mathcal{E}_e = \frac{\mathcal{E}_0}{1 + i\dfrac{\omega_c}{\omega_r}}, \tag{68}$$

with $\omega_c = \dfrac{\sigma_0}{\epsilon}$.

Finally, the average power dissipation P_D is, taking $\sigma \simeq \sigma_0$,

$$P_D = \frac{1}{2} \text{Re } (\sigma \mathcal{E}_e \mathcal{E}_e^*) \rightarrow P_D \simeq \frac{1}{2} \sigma_0 \mathcal{E}_0^2 \frac{1}{1 + \left(\dfrac{\omega_c}{\omega_r}\right)^2}. \tag{69}$$

Equation (69) is the same as Equation (9).

The inclusion of diffusive current flow in the analysis is straightforward. Equation (67) is replaced by

$$\vec{J} = \sigma \vec{\mathcal{E}}_e + D \nabla \rho_i \rightarrow J \simeq \sigma_0 \mathcal{E}_e - iDk\rho_i. \tag{70}$$

Now the continuity equation yields, in place of Equation (66),

$$\rho_i = \frac{k}{\omega_r} J = \frac{k\sigma_0/\omega_r}{1 + i\dfrac{Dk^2}{\omega_r}} \mathcal{E}_e, \tag{71}$$

and the relations in Equation (68) become

$$\mathcal{E}_i = \frac{i\dfrac{\omega_c}{\omega_r}\mathcal{E}_e}{1+i\dfrac{Dk^2}{\omega_r}}; \quad \mathcal{E}_e = \frac{1+i\dfrac{Dk^2}{\omega_r}}{1+i\left[\dfrac{Dk^2+\omega_c}{\omega_r}\right]}. \tag{72}$$

Using Equations (70) and (71), Equation (67) is replaced by

$$J \simeq \sigma_0 \mathcal{E}_e \frac{1}{1+i\dfrac{Dk^2}{\omega_r}}. \tag{73}$$

Finally, Equation (69) is replaced by

$$P_D = \frac{1}{2}\mathrm{Re}\cdot(J\mathcal{E}_e^*) \rightarrow P_D = \frac{1}{2}\sigma_0\mathcal{E}_0{}^2 \frac{1}{1+\left(\dfrac{\omega_D+\omega_c}{\omega_r}\right)^2}; \qquad \omega_0 = Dk^2 \tag{74}$$

and $$\frac{dE}{dt} = P_D\frac{\omega}{\omega_r} = \frac{\epsilon\mathcal{E}_0{}^2\omega}{L}\frac{\omega_r/\omega_c}{\left(1+\dfrac{\omega_D}{\omega_c}\right)^2+\left(\dfrac{\omega_r}{\omega_c}\right)^2}, \tag{75}$$

where we have used $\sigma_0 = \epsilon\omega_c$. (Equation (75) was used as the basis for Equation (18).)

APPENDIX II*—AMPLIFICATION OF LIGHT BY CERENKOV ELECTRONS IN A SOLID

Since electrons drifting faster than the speed of light in a solid (i.e., Cerenkov electrons) can spontaneously emit light, we would also expect that they can coherently amplify light. The calculation presented below explicitly exhibits the expected amplification.

The relevant equations, written in MKS units, are the Maxwell Equations (76)-(79) and the Lorentz force Equation (80);

$$\nabla\cdot\vec{\mathcal{D}} = \epsilon\nabla\cdot\vec{\mathcal{E}} = \rho, \qquad (\vec{\mathcal{D}} = \epsilon\vec{\mathcal{E}}) \tag{76}$$

* The Appendixes to this paper were written by M. A. Lampert.

$$\nabla\cdot\vec{\mathcal{B}} = \mu_0 \nabla\cdot\mathcal{H} = 0, \qquad (\vec{\mathcal{B}} = \mu_0 \vec{\mathcal{H}}) \tag{77}$$

$$\nabla \times \vec{\mathcal{E}} = -\frac{\partial \vec{\mathcal{B}}}{\partial t} = -\mu_0 \frac{\partial \vec{\mathcal{H}}}{\partial t}, \tag{78}$$

$$\nabla \times \vec{\mathcal{H}} = \vec{j} + \frac{\partial \vec{\mathcal{D}}}{\partial t} = \vec{j} + \epsilon \frac{\partial \vec{\mathcal{E}}}{\partial t}, \tag{79}$$

$$m\left(\frac{d\vec{v}}{dt} + \frac{\vec{v}}{\tau}\right) = e(\vec{\mathcal{E}} + \vec{v} \times \vec{\mathcal{B}}) = e(\vec{\mathcal{E}} + \mu_0 \vec{v} \times \vec{\mathcal{H}}). \tag{80}$$

Note that the ϵ in Equation (76) is the dielectric constant of the solid, at the light-wave frequency, in the absence of the conduction electrons. Also note that we have included a collision-induced friction term, in the usual manner, in the Lorentz-force equation.

The current density may be written

$$\vec{j} = en\vec{v} = \sigma\vec{\mathcal{E}}; \qquad \sigma = en\mu \text{ with } \mu = \vec{v}/\vec{\mathcal{E}}. \tag{81}$$

Finally, we have

$$\frac{d}{dt} = \frac{\partial}{\partial t} + \vec{v}\cdot\nabla. \tag{82}$$

In the absence of electron drift, the solid will support a uniform, transverse plane wave of the form

$$\vec{\mathcal{E}} = \hat{x}\mathcal{E} \exp i(\omega t - kz), \tag{83}$$

$$\vec{\mathcal{H}} = \hat{y}\mathcal{H} \exp i(\omega t - kz). \tag{84}$$

We now look for the same kind of solution in the presence of electron drift. Equation (76) is satisfied with $\rho \equiv 0$. Equation (77) is automatically satisfied. Equations (78) and (79) become, respectively,

$$k\mathcal{E} = \omega\mu_0\mathcal{H}, \tag{85}$$

$$k\mathcal{H} = (\omega\epsilon - i\sigma)\mathcal{E}. \tag{86}$$

Combining the two equations we get

$$k^2 = k_0{}^2 \left(1 - i\frac{\sigma}{\omega t}\right), \quad k_0{}^2 = \omega^2 \mu_0 \epsilon, \tag{87}$$

where k_0 is the propagation constant in the solid in the absence of conduction electrons.

Writing $\vec{v} = \vec{v_0} + v\hat{x} \exp i(\omega t - kz)$ and linearizing Equations (82) and (80), i.e., neglecting the product of a-c amplitudes, we obtain

$$\left[i(\omega - kv_0) + \frac{1}{\tau}\right] \vec{v} = \frac{e}{m} [\mathcal{E} - \mu_0 v_0 \mathcal{H}]\hat{x}. \tag{88}$$

Finally, substituting for $\mu_0 \mathcal{H}$ from Equation (85) into Equation (88) and using Equation (81) we obtain

$$\sigma = \sigma_0 \frac{1 - \dfrac{kv_0}{\omega}}{1 + c\omega\tau\left(1 - \dfrac{kv_0}{\omega_p}\right)}, \quad \sigma_0 = \frac{e^2 n_0 \tau}{m}, \tag{89}$$

where σ_0 is the d-c conductivity. Note that for nondrifting electrons, $v_0 = 0$, Equation (89) reduces to the well-known a-c conductivity.

Substitution of Equation (89) into Equation (87) gives the dispersion relation for the propagation. It is convenient now to separate k into its real and imaginary parts;

$$k = k_1 - ik_2. \tag{90}$$

Propagation is now described by

$$\vec{\mathcal{E}} = \hat{x}\mathcal{E}_0 \exp i(\omega t - k_1 z) \exp(-k_2 z) \tag{91}$$

and likewise for $\vec{\mathcal{H}}$. Thus positive k_2 corresponds to attenuation, negative k_2 to growth.

Further, Equation (89) may be rewritten as

$$\frac{\sigma}{\sigma_0} = \frac{\left(1 - \frac{k_1 v_0}{\omega_p}\right) + i\,\frac{k_2 v_0}{\omega}}{(1 - k_2 v_0 \tau) + i\tau(\omega - k_1 v_0)}$$

$$= \frac{1 - \frac{k_1 v_0}{\omega}}{(1 - k_2 v_0\tau)^2 + (\omega\tau)^2\left(1 - \frac{k_1 v_0}{\omega}\right)^2} + i\,\frac{\frac{k_2 v_0}{\omega}(1 - k_2 v_0\tau) - \omega\tau\left(1 - \frac{k_1 v_0}{\omega}\right)}{(1 - k_2 v_0\tau)^2 + (\omega\tau)^2\left(1 - \frac{k_1 v_0}{\omega}\right)^2}. \tag{92}$$

Separation of the dispersion relation into its real and imaginary parts gives the two equations:

$$\frac{k_1^2 - k_2^2}{k_0^2} = 1 + \frac{1}{\omega t_r}\,\frac{\frac{k_2 v_0}{\omega}(1 - k_2 v_0\tau) - \omega\tau\left(1 - \frac{k_1 v_0}{\omega}\right)^2}{(1 - k_2 v_0\tau)^2 + (\omega\tau)^2\left(1 - \frac{k_1 v_0}{\omega}\right)^2} \tag{93}$$

$$\frac{k_1 k_2}{k_0^2} = \frac{1}{2\omega t_r}\,\frac{1 - \frac{k_1 v_0}{\omega}}{(1 - k_2 v_0\tau)^2 + (\omega\tau)^2\left(1 - \frac{k_1 v_0}{\omega}\right)^2}; \qquad t_r = \frac{\epsilon}{\sigma_0} \tag{94}$$

where t_r is the well-known dielectric relaxation time.

Equations (93) and (94) are a pair of coupled algebraic equations in the unknowns k_1 and k_2. Letting $k_1 = 2\pi/\lambda$, we see from Equation (91) that the attenuation (growth) per wavelength is given by $\exp(-k_2\lambda) = \exp(-2\pi k_2/k_1)$. The simplest case to discuss is that of weak coupling, i.e., relatively slow attenuation (growth), $k_2 \ll k_1$, and a propagation characteristic very close to that in the solid in the absence of conduction electrons (second term on the right-hand side of Equation (93) small compared to unity). Then Equations (93) and (94) reduce to

$$k_1 = k_0 = \omega\sqrt{\mu_0\epsilon}, \quad v_w = \frac{\omega}{k_1} = \frac{1}{\sqrt{\mu_0\epsilon}} = \frac{c}{\sqrt{\epsilon/\epsilon_0}}, \tag{95}$$

$$\frac{k_2}{k_1} = \frac{1}{2\omega t_r} \frac{1 - \dfrac{v_0}{v_w}}{\left(1 - \omega\tau \dfrac{k_2 v_0}{k_1 v_w}\right)^2 + (\omega\tau)^2 \left(1 \dfrac{v_0}{v_w}\right)^2} . \tag{96}$$

It is immediately clear from Equation (96) that for

1. $v_1 = v_w$, $k_2 = 0$ and there is no interaction of the light wave with the synchronously drifting electrons. This is expected since, for the unperturbed light wave, the Lorentz force $\vec{\mathcal{E}} + \vec{v}_0 \times \vec{\mathcal{B}} = \vec{\mathcal{E}} + \vec{v}_w \times \vec{\mathcal{B}} = 0$.

2. $v_0 > v_w$, $k_2 < 0$ and the light wave is amplified. In particular, for $v_0 = 2v_w$ Equation (96) becomes approximately

$$k_2 = \frac{\sqrt{\epsilon/\epsilon_0}}{2ct_r(\omega\tau)^2}, \tag{97}$$

which is numerically just the expression for absorption of light by free carriers but with reversed sign, indicating amplification of the light wave. For larger values of v_0, k_2 decreases as $1/v_0$. Hence the behavior for light waves completely parallels that for acoustic waves.

Reprinted from *RCA Review*, 27 (1966) 600,

Chapter IV

Energy losses by hot electrons

BY

ALBERT ROSE

RCA Laboratories
Princeton, N. J.

Editor's Note: This is the second part of a several part paper. Part I appeared in the March 1966 issue of *RCA Review* (Vol. XXVII, No. 1, p 98).

Summary—*The rates of energy loss by hot electrons to phonons and to electronic excitations are derived from an elementary model. All of the energy-loss expressions have essentially the same form. The various channels for loss of energy are characterized by a coupling constant which, in the case of phonons, is shown to be the ratio of electrical to total energy for the corresponding macroscopic sound waves. It is the same coupling constant which, in Part I, was shown to characterize the various acousto-electric effects. Quantum effects are introduced as constraints on the classical model. Finally, the rates of energy loss are related to the acousto-electric effects by a heuristic argument.*

1. INTRODUCTION

WHEN an electric field is applied to a solid, the free carriers are heated relative to the lattice. The degree to which the carriers are heated depends on the rate at which they can lose energy to the lattice. Since the rate of energy loss is rather insensitive to the lattice temperature, we will assume a zero-temperature lattice.

The gamut of energy-loss mechanisms is shown in Table I. At the low end of the scale, electrons whose energies lie in the approximate range 1 to 25 millivolts lose energy to acoustic phonons. The coupling to the acoustic phonons may be via a deformation potential or a piezoelectric field. At about 25 millivolts the electrons begin to lose energy

Table I

Form of Radiated Energy	Approximate Threshold Excitation Energy Electron Volts	Analyzed By	Year
Acoustic Phonons	10^{-3}		
Deformation-Potential		Seitz[1]	1949
Piezoelectric		Tsu[2]	1964
Intervalley Scattering		Herring[3]	1955
Optical Phonons	10^{-2}		
Polar		Fröhlich[4]	1939
Nonpolar		Conwell[5]	1959
Impact Ionization	E_{gap}	Motizuki and Sparks[6]	1964
Plasmons	10	Bohn and Pines[7]	1953
Excitation of X-Ray Levels	10^{2}	Bohr[8]	1913
		Bethe[9]	1929
Cerenkov Radiation	10^{4}	Frank and Tamm[10]	1937

[1] F. Seitz, "On the Theory of Electron Multiplication in Crystals," *Phys. Rev.*, Vol. 76, p. 1376, Nov. 1, 1949.

[2] R. Tsu, "Phonon Radiation by Uniformly Moving Charged Particles in Piezoelectric Solids," *Jour. Appl. Phys.*, Vol. 35, p. 125, Jan. 1964. See also H. J. G. Meijer and D. Polder, "Notes on Polar Scattering of Conduction Electrons in Regular Crystals," *Physica*, Vol. 19, p. 255, 1953.

[3] C. Herring, "Transport Properties of a Many-Valley Semiconductor," *Bell Syst. Tech. Jour.*, Vol. 34, p. 237, March 1955.

[4] H. Fröhlich, "On the Theory of Dielectric Breakdown in Solids," *Proc. Roy. Soc.* (London), Vol. A188, p. 521, 25 Feb. 1947; H. Fröhlich, "Theory of Electrical Breakdown in Ionic Crystals," *Proc. Roy. Soc.*, Vol. A160, p. 230, 18 May 1937.

[5] E. M. Conwell, "Relative Energy Loss to Optical and Acoustic Modes of Electrons in Avalanche Breakdown in Ge," *Phys. Rev.*, Vol. 135, p. A1138, 17 Aug. 1964.

[6] K. Motizuki and M. Sparks, "Range of Excited Electrons and Holes in Metals and Semiconductors," *Jour. Phys. Soc. Japan*, Vol. 19, p. 486, April 1954.

[7] D. Pines and D. Bohm, "A Collective Description of Electron Interactions; II. Collective vs. Individual Particle Aspects of the Interactions," *Phys. Rev.*, Vol. 85, p. 338, 15 Jan. 1952; and "A Collective Description of Electron Interactions: III. Coulomb Interactions in a Degenerate Electron Gas," *Phys. Rev.*, Vol. 92, p. 609, Nov. 1, 1953.

[8] N. Bohr, "Decrease of Speed of Electrified Particles in Passing Through Matter," *Phil. Mag.*, Vol. 25, p. 10, Jan. 1913 and Vol. 30, p. 581, Oct. 1915.

[9] H. Bethe, *Ann. Phys.*, Vol. 16, p. 285, 1933; *Handbuch der Physik* (Geiger and Scheel, Eds.), Vol. 24, Part I, Springer, Berlin, 1933; (see also Reference (11) for critical historical review).

[10] I. Frank and I. Tamm, "Coherent Visible Radiation from Fast Electrons Passing Through Matter," *Comptes Rendus (Doklady) de l'Acad. des Sci.*, USSR, Vol. 14, p. 109, 1937.

[11] J. D. Jackson, *Classical Electrodynamics*, John Wiley and Sons, New York, 1963.

to optical phonons, both polar and nonpolar. For energies of the order of volts, they can lose energy by impact ionization across the forbidden gap. In the range of about ten to twenty volts they begin to excite plasmon oscillations. At the level of hundreds of volts they lose energy to the deeper-lying atomic levels. Finally, at speeds approaching the velocity of light, the electrons emit Cerenkov radiation. All of the above energies represent threshold values. High-velocity electrons lose energy to all of the above mechanisms simultaneously.

The rates of energy loss to each of the channels listed have been computed and have appeared in the literature. Some of the names and dates associated with these publications are listed in Table I. It is interesting to note that the dates are more or less uniformly distributed over the last fifty years. The methods of analysis are almost as varied as the number of contributors. Moreover, most of the analyses, particularly those for energy loss to phonons, are carried out in momentum space. The following discussion is an attempt to re-state the arguments in "real space" and in terms of an elementary model. The results have only an order-of-magnitude validity, consistent with the use of the uncertainty principle.

An outline of the argument is presented in the next four sections. This is followed by a more detailed consideration of each of the modes of energy loss. The central concept is that of an electron polarizing the surrounding medium to form an energy well. The moving electron loses energy in the form of a wake, or trail, of energy wells.

2. Dimensional Argument

Figure 1 gives a dimensional argument for the rate of loss of energy by a moving particle to a system with which it interacts. In Figure 1 a particle moves with velocity v past a series of elements of dimension d. The particle repels each element with a force such that, for a stationary particle, an energy E_w is stored in the "compressed spring" of the element. The frequency of vibration of each element is denoted by ω.

By inspection we can write the maximum rate of loss of energy by the particle:

$$\left.\frac{dE}{dt}\right]_{max} \approx E_w \omega. \qquad (1)$$

At the assumed velocity ωd, the particle spends a time ω^{-1} opposite each element and deflects the element almost as far as it would if the particle were stationary. While the element is in the deflected state the particle moves on to the next element leaving behind an energy of almost E_w per element. Since, by assumption, the particle traverses ω elements per second, the rate of energy loss is that given by Equation (1).

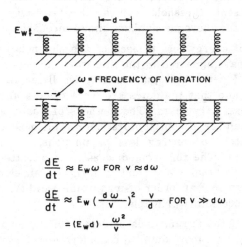

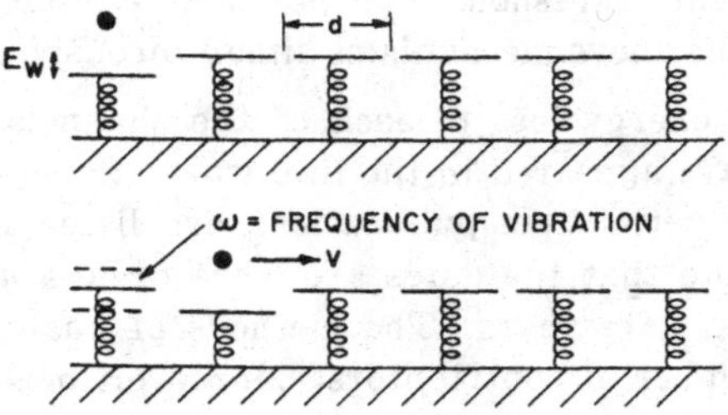

$$\frac{dE}{dt} \approx E_w \omega \text{ FOR } v \approx d\omega$$

$$\frac{dE}{dt} \approx E_w \left(\frac{d\,\omega}{v}\right)^2 \frac{v}{d} \text{ FOR } v \gg d\omega$$

$$= (E_w d)\frac{\omega^2}{v}$$

Fig. 1—Model for computing rate of energy loss by a moving particle.

A quick confirmation of Equation (1) is obtained by expressing the maximum rate of loss of energy to polar optical phonons as the product of the polaron energy $\alpha\hbar\omega$ and the optical-phonon frequency ω:

$$\left.\frac{dE}{dt}\right]_{max} = (\alpha\,\hbar\,\omega)\,\omega$$

$$= \frac{e^2}{\bar{\epsilon}}\left(\frac{m}{\hbar^3\omega}\right)^{1/2}\hbar\omega^2 = \frac{e^2\omega^2}{\bar{\epsilon}v} \quad \text{for } \tfrac{1}{2}mv^2 \approx \hbar\omega, \qquad (2)$$

where $\bar{\epsilon}^{-1} \equiv \epsilon_\infty^{-1} - \epsilon_0^{-1}$. Equation (2) matches the result obtained by Fröhlich[4] and by Callen[12] using perturbation theory. In this argument we have, of course, borrowed the concept and magnitude of a polaron energy well from the results of perturbation theory. We will show later a semiclassical method of computing E_w.

Meantime, returning to Figure 1, we observe, again almost by

[12] H. B. Callen, "Electric Breakdown in Ionic Crystals," *Phys. Rev.*, Vol. 76, p. 1394, Nov. 1, 1949.

inspection, that the rate of energy loss for $v > \omega d$ is

$$\frac{dE}{dt} = E_w \left(\frac{\omega d}{v}\right)^2 \frac{v}{d} = E_w d \frac{\omega^2}{v}. \tag{3}$$

The factor $(\omega d)/v$ is equal to $\omega\tau$, where τ is the transit time of the particle past one of the elements of the array. Equation (3) reduces to Equation (1) for $\omega\tau = 1$. For $v > \omega d$, $\omega\tau < 1$ so that the particle acts on each element with an impulsive force. The momentum imparted is proportional to τ. Hence, the energy imparted is proportional to τ^2. The energy imparted per element is therefore less than E_w by the factor $(\omega\tau)^2$ where $\omega\tau < 1$. The last factor, v/d, is simply the number of elements traversed per second.

Equation (3) is the essence of the "real-space" argument for energy losses by fast electrons. In order to apply Equation (3) to electrons, we need to determine in each case the value of E_w, the energy well that a stationary electron digs in the surrounding medium or, more definitely, in the mode of vibration of the medium through which the energy loss is being computed.

For those cases where the electron couples to and does work on the medium via its coulomb field, we can write

$$E_w = \beta \frac{e^2}{Kd}, \tag{4}$$

where β is, by definition, the fraction of the available coulomb energy that is used to form an energy well and K is the real part of the dielectric constant for frequencies higher than that of the emitted radiation. For example, if the electron is emitting energy to polar optical phonons, K is equal to ϵ_∞, the high-frequency, or electronic, part of the dielectric constant. In brief, the coulomb energy available for doing work on the ionic part of the lattice is reduced from its vacuum value by the electronic polarization factor ϵ_∞.

Combination of Equations (3) and (4) yields

$$\frac{dE}{dt} \approx \beta \frac{e^2\omega^2}{Kv}. \tag{5}$$

This is the form in which the various rates of energy loss, taken from the published literature, have been recast in Table II. In the case of coupling by deformation potential, a coulomb energy of $e^2/(Kd)$, d

Table II—Time Rates of Energy Loss (dE/dt) by Electrons of Velocity v.

Phenomenon	dE/dt	Source
Polar Optical Phonons	$\left[\frac{\epsilon_o - \epsilon_x}{\epsilon_o}\right]\frac{e^2\omega^2}{\epsilon_x v}\ln\left(\frac{2m\,v^2}{\hbar\omega}\right)$ Note: $\frac{1}{2}mv^2 > \hbar\omega$	Fröhlich[4] Callen[12]
Piezoelectric Phonons	$\frac{\pi}{4}\left[\frac{\epsilon_p^2}{KC}\right]\frac{e^2\omega^2}{Kv}$ Note: $2mv = \hbar\omega/v_s$	Tsu[2]
Acoustic Phonons	$\frac{1}{4}\left[\frac{B^2\omega^2K}{4\pi e^2\rho v_s^4}\right]\frac{e^2\omega^2}{Kv}$ Note: $2mv = \hbar\omega/v_s$	Seitz[1] Conwell[5]
Nonpolar Optical Phonons	$\frac{1}{2}\left[\frac{\pi K D^2}{\rho e^2\omega^2\lambda^2}\right]\frac{e^2\omega^2}{Kv}$ Note: $\frac{1}{2}mv^2 > \hbar\omega$ and $\frac{\lambda}{2\pi} = \frac{\hbar}{2mv}$	Conwell[5]
X-Ray Levels	$\left[\frac{\omega_p^2}{\omega_e^2}\right]\frac{e^2\omega_e^2}{v}\ln\left(\frac{2mv^2}{\hbar\omega_e}\right)$ Note: $\omega_p^2 = \frac{4\pi n e^2}{m} \ll \omega_e^2$; $\hbar\omega_e$ = Excitation energy of x-ray levels; $\frac{1}{2}mv^2 > \hbar\omega_e$	Bohr[8] Bethe[9]
Plasma	$\left[1\right]\frac{e^2\omega^2}{v}\ln\left(\frac{2mv^2}{\hbar\omega}\right)$ Note: ω = plasma frequency and $\frac{1}{2}mv^2 > \hbar\omega$	Bohm and Pines[7]
Cerenkov	$\left[1 - \frac{1}{\epsilon_r}\right]\frac{e^2\omega^2}{v}$ Note: $v = c$ = velocity of light in vacuum	See, e.g., Schiff[13]

[13] L. I. Schiff, *Quantum Mechanics*, p. 271, McGraw-Hill Book Co., New York, 1955.

Definitions for Table II

ϵ_0 = low-frequency dielectric constant

ϵ_∞ = high-frequency (optical) dielectric constant

K = dielectric constant

ϵ_p = piezoelectric constant

C = elastic modulus (dynes/cm^2)

ρ = density (grams/cm^3)

v_s = phase velocity of sound

v = velocity of electron

ω = angular frequency of radiation

B = deformation potential (electron volts in ergs/unit strain)

D = optical deformation potential (electron volts in ergs per centimeter relative shift of sublattices)

m = effective mass of electrons

being an uncertainty diameter, was assumed. This procedure may or may not have physical significance. It does allow ready comparison with the other rates of energy loss.

3. Coupling Constant

We need at this point to clarify the meaning of β. We consider first the loss to phonons, and write

$$\frac{1}{2}\left[\mathcal{E}_c^2 - (\mathcal{E}_c - \mathcal{E}_p)^2\right] = \mathcal{E}_p^2 + E_{elastic}, \qquad (6)$$

where $\mathcal{E}_c$ is the coulomb field of the electron, $\mathcal{E}_p$ is the (opposing) polarization field induced in the medium, and $E_{elastic}$ is the elastic energy stored in the medium (since a polarization field is accompanied by a mechanical strain or distortion of the medium). We have omitted factors of 8π and dielectric constants for simplicity and because they cancel out in the ratios obtained below. The left-hand side of Equation (6) is the coulomb energy per unit volume *available* for doing work in the medium. It is the difference between the electrostatic energy in the field of the electron before and after polarizing the medium. The factor 1/2 takes into account that the efficiency at which the stored energy can do useful work is, at most, 50%. The right-hand side of

Equation (6) is the energy imparted to the medium and consists of an electrical part $\mathcal{E}_p^2$ and an elastic or mechanical part $E_{elastic}$. An elementary example is that a strain in a piezoelectric crystal has an elastic energy that is distinct from the electrical energy due to the piezoelectric field.

For $\mathcal{E}_p \ll \mathcal{E}_c$, Equation (6) can be written

$$\mathcal{E}_c\mathcal{E}_p = \mathcal{E}_p^2 + E_{elastic}. \tag{7}$$

Using Equation (7), we write for the fraction of coulomb energy used in forming the energy well,

$$\beta \equiv \underbrace{\frac{\mathcal{E}_c\mathcal{E}_p}{\mathcal{E}_c^2}}_{\text{Fraction of coulomb energy used in forming energy well.}} = \frac{\mathcal{E}_p^2}{\mathcal{E}_c\mathcal{E}_p} = \Bigg) \underbrace{\frac{\mathcal{E}_p^2}{\mathcal{E}_p^2 + E_{elastic}}}_{\left.\frac{\text{electrical energy}}{\text{total energy}}\right|_{\text{phonon}}}. \Bigg) \tag{8}$$

Equation (8) says that our general definition of β is equivalent to taking the ratio of electrical to total energy for the various phonons.

The values of β for the phonons in Table II are shown in square brackets. They were computed in Chap. III for macroscopic sound waves and were used (see Table III) to characterize the various acoustoelectric effects. For polar optical waves, we used the Lyddane–Sachs–Teller relation for the ratio of frequencies of transverse and longitudinal modes and combined it with the fact that transverse modes have only elastic energy while the longitudinal modes have the same elastic energy plus an electric field energy due to polarization. For piezoelectric waves, the ratio of electrical to elastic energy is given by the square of the well-known electromechanical coupling constant. Since the electrical energy is small compared with the elastic energy, this ratio is a good approximation to the ratio of electrical to total energy. For the acoustic and nonpolar optical waves coupled by deformation potential, we computed the electrical energy as if the slope of the band edges due to the deformation potential were an actual field.

The values of β for the electronic excitations are close to unity for both plasmons and Cerenkov radiation, since the coulomb field of the electron is substantially canceled by electronic polarization and is therefore substantially completely used in digging an energy well. In the case of the deeper lying atomic levels (this is essentially the

* See *RCA Review*, March 1966, pp. 98-139.

problem treated classically by Bohr)[8], the tightly bound electrons can only partially cancel the coulomb field of the impinging electron since their contribution to the dielectric constant is near unity, namely, $1 + (\omega_p^2/\omega_e^2)$. Here $\hbar\omega_e$ is their binding energy, $\omega_p^2 = 4\pi ne^2/m$, and $\omega_p^2 \ll \omega_e^2$. It is this small value of dielectric constant that leads to the value of $\beta = \omega_p^2/\omega_e^2$ in Table II.

4. Inner Radius of Coulomb Field

We have thus far given the dimensional form (Equations (3) and (5)) for the rates of energy loss and have defined the meaning of β as the fraction of available coulomb energy used in digging an energy well. What remains to be clarified further is the phrase "available coulomb energy." We have pointed out, for example, that in the case of loss to polar optical phonons the available coulomb energy is the vacuum field energy reduced by the high-frequency or electronic part of the dielectric constant. In the case of loss to electronic excitations, it is the vacuum field energy itself. What we have not yet defined is the inner radial limit of this coulomb field.

The inner radius is set by the uncertainty relation. This means, for example, that the radius of the smallest wave packet that can be formed for an electron whose maximum momentum is mv is of the order of $\hbar/mv$. Within this uncertainty radius the electronic charge is taken to be more or less uniformly distributed.

For energy loss to those phonons where the coupling is a real macroscopic electric field (polar optical and piezoelectric) the coulomb field energy extends from the uncertainty radius outward. Figure 2 shows schematically how this coulomb energy can, in the case of polar optical phonons where ω and β are constants independent of the λ of the radiated energy, be divided into a series of equivalent shells (in Figure 2, $\beta = K = 1$). The rate of energy loss for each of these shells is $\beta(e^2\omega^2)/(Kv)$. Summing up over the shells yields the following

$$\frac{dE}{dt} \approx \beta\frac{e^2\omega^2}{Kv}\ln\frac{r_o}{r_i}$$

$$\approx \beta\frac{e^2\omega^2}{Kv}\ln\frac{mv^2}{\hbar w}, \qquad (9)$$

so that the effect of the summation is to introduce a logarithmic factor. The outer radius is defined by $2r_o = v\omega^{-1}$ and is that radius beyond which the medium is slowly and reversibly polarized and depolarized

Table III—Time Constant (ω_a) for Rate of Change of Energy Density in Sound Waves

Phenomenon	ω_a	Source
Polar Optical	$\left[\dfrac{\epsilon_0 - \epsilon_\infty}{\epsilon_0}\right] \omega f(\omega_r, \omega_c, \omega_D)$	Gunn[14] Woodruff[15]
Piezoelectric Acoustic	$\left[\dfrac{\epsilon_p^2}{KC}\right] \omega f(\omega_r, \omega_c, \omega_D)$	Hutson, McFee and White[16]
Deformation Potential	$\left[\dfrac{KB^2\omega^2}{4\pi\rho e^2 v_s^4}\right] \omega f(\omega_r, \omega_c, \omega_D)$	Parmenter[17] Weinreich[18]
Nonpolar Optical	$\left[\dfrac{\pi KD^2}{\rho e^2 w^2 \lambda^2}\right] \omega f(\omega_r, \omega_c, \omega_D)$	
Intervalley Deformation Potential	$\left[\dfrac{nB^2}{\rho v_s^2 kT}\right] \omega f(\omega_r, \omega_c)$	Weinreich[19] Pomerantz[20]
Metals	$\left[1\right] \omega f(\omega_r, \omega_c)$ or $\omega^2\tau_c$ for $\omega_r = \omega$ and $\omega\tau_c \ll 1$	Pippard[21] Levy[22]

Note that these expressions are valid for the mean free path of the electrons less than the wavelength of the sound wave.

[14] J. B. Gunn, "Traveling-Wave Interaction between the Optical Modes of a Polar Lattice and a Stream of Charge Carriers, "*Physics Letters*, Vol. 4, p. 194, 1 April 1963.

[15] T. O. Woodruff, "Interaction of Waves of Current and Polarization," *Phys. Rev.*, Vol. 132, p. 679, 15 Oct. 1963.

[16] A. R. Hutson, J. H. McFee, and D. L. White, "Ultrasonic Amplification in CdS," *Phys. Rev. Letters*, Vol. 7, p. 237, Sept. 15, 1961.

[17] R. H. Parmenter, "The Acousto-Electric Effect," *Phys.* Rev., Vol. 89, p. 990, March 1, 1953.

[18] G. Weinreich, "Ultrasonic Attenuation by Free Carriers in Germanium," *Phys. Rev.*, Vol. 107, p. 317, July 1, 1957.

[19] G. Weinreich, T. M. Sanders, Jr., and H. G. White, "Acoustoelectric Effect in n-Type Germanium," *Phys. Rev.*, Vol. 114, p. 33, April 1, 1959.

[20] M. Pomerantz, "Amplification of Microwave Phonons in Germanium," *Phys. Rev. Letters*, Vol. 13, p. 308, 31 Aug. 1964.

Definitions for Table III

$$\omega_a = \frac{1}{E}\frac{dE}{dt};\ E = \text{energy density of acoustic wave}$$

$$f(\omega_r, \omega_c, \omega_D) = \frac{\omega_r/\omega_c}{\left(1+\frac{\omega_D}{\omega_c}\right)^2 + \left(\frac{\omega_r}{\omega_c}\right)^2}$$

$$f(\omega_r, \omega_c) = \frac{\omega_r/\omega_c}{1+(\omega_r/\omega_c)^2}$$

$$\omega_r = \frac{v_d - v_s}{v_s}\,\omega$$

$$\omega_D = \frac{4\pi^2 kT\mu}{e\lambda^2}$$

$\omega_c = 4\pi\sigma/K$, or $\omega_c = \tau_c^{-1}$ for metals and intervalley deformation where τ_c is the time for an electron to come to thermal equilibrium

v_d = drift velocity of electrons

v_s = phase velocity of sound

ϵ_o = low frequency dielectric constant

ϵ_∞ = high frequency (optical) dielectric constant

K = dielectric constant

ϵ_p = piezoelectric constant

C = elastic modulus

B = deformation potential (electron volts in ergs/unit strain)

D = optical deformation potential (electron volts in ergs per centimeter relative shift of sublattices)

---

[21] A. B. Pippard, "Ultrasonic Attenuation in Metals," *Phil. Mag.*, Vol. 46, p. 1104, Oct. 1955.

[22] M. Levy, "Ultrasonic Attenuation in Superconductors for $ql < 1$," *Phys. Rev.*, Vol. 131, p. 1497, 15 Aug. 1963.

without significant energy loss. The same argument holds for energy loss to plasmons, in which case $\beta = K = 1$.

In the case of acoustic phonons coupled by piezoelectric fields the major loss is due to the innermost shell, and the logarithmic factor becomes unity. The outer shells by virtue of their larger dimensions excite longer wavelength radiations and hence are associated with

$$\left.\frac{dE}{dt}\right|_{\frac{r_0}{n}} = (E_w d)\frac{\omega^2}{v}$$

$$= \left(\frac{ne^2}{2r_0}\cdot\frac{2r_0}{n}\right)\frac{\omega^2}{v} \; ; 2r_0 = \frac{v}{\omega}$$

$$= \frac{e^2\,\omega^2}{v}$$

$$\sum_n \frac{dE}{dt} \approx \frac{e^2\,\omega^2}{v}\,\ell n\,\frac{r_0}{r_i}$$

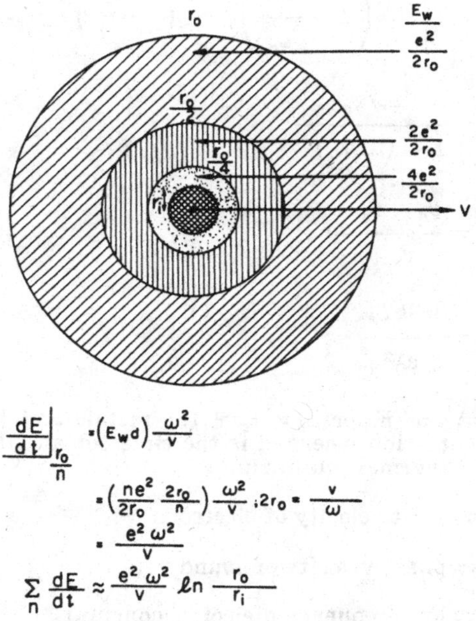

Fig. 2—Resolution of the rate of energy loss into a series of equivalent shells ($\beta = K = 1$ and $\omega = $ constant).

smaller values of ω. Since ω enters in as ω^2, the contributions of the outer shells are small in comparison with the innermost shell.

The energy loss to phonons coupled by a deformation potential can be cast in the same form as the other interactions in which the electric field plays a clearly identifiable role. To do so, the available coulomb energy is taken to be of the order of $e^2/(Kd)$ where d is the uncertainty diameter of the electron. While it is true that the concept of the deformation potential is not related in any direct or simple way to the coulomb field and energy, it may be more than accident that this particular value of the available coulomb energy is needed to satisfy the formalism.

The energy well for deformation potential coupling can also be obtained directly from balancing the interaction energy gained by an electron in a strained lattice with the energy needed to produce the strain;

$$\frac{1}{2} BS \approx \frac{1}{2} CS^2 \left(\frac{4\pi}{3} r^3 \right) .$$

Here B is the deformation potential, S the strain, C an elastic constant, and r the uncertainty radius. The energy well is, therefore,

$$\frac{1}{2} BS \approx \frac{3B^2}{8\pi C r^3}. \tag{10}$$

This value is inserted into Equation (3) to obtain the rate of energy loss.

Note that the inner radius enters only into the rather insensitive logarithmic factor for polar optical phonons. For the other phonons the inner radius has a first-order effect in determining the smallest wavelength or largest frequency (ω) phonon that can be emitted.

5. Other Problems

Equation (5) has also been applied to the rate of loss of energy to phonons via intervalley scattering; the computation is similar to that for loss to intravalley phonons coupled by a deformation potential. The difference is that the momentum required to emit the high-momentum phonons connecting the two valleys is obtained from the change in crystal momentum when the electron makes a transition between valleys, as well as from the local momentum of the electron in its valley.

Equation (5) is a useful approximation for computing energy loss via impact ionization across a forbidden gap, particularly if the forbidden gap is larger than the width of the valence band. The problem then approaches that for excitation of deep-lying atomic levels having a spread in excitation energies.

6. Polar Optical Phonons

The rate of energy loss to polar optical phonons* is given by **Equation (9) combined with the value for β taken from Chap. III:**

* See H. Fröhlich and N. Pelzer. "Polarization of Dielectrics by Slow-Particles," E.R.A. Report No. L/T184(1948), for classical treatment of this problem.

* See *RCA Review*, March 1966, pp. 98-139.

$$\frac{dE}{dt} = \left[\frac{\epsilon_0 - \epsilon_\infty}{\epsilon_0}\right] \frac{e^2\omega^2}{\epsilon_\infty v} \ln \frac{mv^2}{\hbar\omega}. \tag{11}$$

According to Callen's[12] result, the logarithmic factor should have a factor of 2 in the numerator. Equation (11) is valid for $mv^2/2 > \hbar\omega$. Callen's factor is

$$\ln \frac{1 + \left(1 - \frac{\hbar\omega}{E}\right)^{1/2}}{1 - \left(1 - \frac{\hbar\omega}{E}\right)^{1/2}},$$

where $E = mv^2/2$. At $E = 2\hbar\omega$, this factor is already within 20% of $\ln(2mv^2/\hbar\omega)$.

It is instructive to also obtain Equation (11) directly from Equation (3), the primitive equation for rate of energy loss. The coulomb energy in a spherical shell of radius r and thickness dr is, in general,

$$dE = \frac{e^2}{Kr^2}\, dr. \tag{12}$$

Consider an electron moving slowly enough to polarize both the ionic and electronic parts of the surrounding lattice. Equation (12) then reads

$$dE = \frac{e^2}{\epsilon_0 r^2}\, dr. \tag{13}$$

Now let the electron move fast enough to polarize only the electronic part. Equation (12) becomes

$$dE = \frac{e^2}{\epsilon_\infty r^2}\, dr. \tag{14}$$

The difference between Equations (13) and (14) is then the energy per unit volume assignable to the polar optical modes. It is the energy well in a shell of diameter $2r$ and thickness dr. We insert this difference in Equation (3) for $d(E_w)$ and obtain

$$\frac{dE}{dt} = \int_{r_i}^{r_0} 2r \frac{\omega^2}{v} d(E_w)$$

$$= \frac{\epsilon_0 - \epsilon_\infty}{\epsilon_0 \epsilon_\infty} \frac{e^2 \omega^2}{v} \int_{r_i}^{r_0} \frac{dr}{r}$$

$$= \left[\frac{\epsilon_0 - \epsilon_\infty}{\epsilon_0} \right] \frac{e^2 \omega^2}{\epsilon_\infty v} \ln \frac{r_0}{r_i}, \tag{15}$$

which is the same result as Equation (11).

The maximum rate of loss of energy occurs near $mv^2/2 = \hbar\omega$ and is, as pointed out in Equation (2), the product of the polaron energy well $\alpha\hbar\omega$ and the optical frequency ω.

7. Piezoelectric Acoustic Phonons

The rate of energy loss to acoustic phonons coupled by a piezoelectric field is, from Equation (5) with $\beta = \epsilon_p^2/(KC)$;

$$\frac{dE}{dt} \approx \frac{\epsilon_p^2}{K^2 C} \frac{e^2 \omega^2}{v}. \tag{16}$$

This result is within a factor of 2 of a quantum mechanical calculation by perturbation theory,* namely,

$$\frac{dE}{dt} = \frac{2\epsilon_p^2}{K^2 C} \frac{e^2 \omega^2}{v}. \tag{17}$$

The Cerenkov-type calculation by Tsu[2] matches Equation (17) except for a numerical constant of order unity.

The highest frequency phonon that can be emitted is given by

* G. D. Whitfield, private communication.

the condition for conservation of momentum†;

$$\frac{\hbar w}{v_s} = 2mv. \tag{18}$$

Equation (18) inserted in Equation (17) yields

$$\frac{dE}{dt} = \frac{2}{C}\left(\frac{2\epsilon_p evm}{\hbar K}\right)^2 v. \tag{19}$$

8. Deformation-Potential Acoustic Phonons

The rate of energy loss to acoustic phonons coupled by a deformation potential is most readily calculated using the primitive Equation (3). The expression for the energy well from Equation (10) is inserted into Equation (3) to obtain

$$\frac{dE}{dt} = \frac{3B^2\,\omega^2}{4\pi\,C\,r^2\,v}. \tag{20}$$

We use the relations‡

$$r \approx \frac{\hbar}{mv} \tag{21}$$

$$\frac{\hbar\omega}{v_s} = 2mv \tag{22}$$

and

$$C = \rho v_s^2 \tag{23}$$

to convert Equation (20) into

$$\frac{dE}{dt} = \frac{3}{\pi}\frac{B^2\,m^4}{\rho\hbar^4}v^3. \tag{24}$$

This is a factor of 3 larger than the result obtained by Seitz[1] and by Conwell[5] from perturbation theory.#

†‡# When the allowed transitions are weighted by a density of states factor, a more realistic value for the right hand side of Equations (18) and (22) is between mv and $2mv$. This refinement brings Equation (24) into substantial agreement with the literature and converts the numerical coefficients in Table II to values close to unity.

Equation (24) can also be obtained using Equation (5) and the value of β given by Equation (37) in Part I†;

$$\beta = \frac{B^2 \omega^2 K}{4\pi e^2 \rho v_s^4}. \tag{25}$$

Substituting the expression for β into Equation (5) yields

$$\frac{dE}{dt} = \beta \frac{e^2 \omega^2}{Kv}$$

$$= \frac{B^2 \omega^4}{4\pi \rho v_s^4 v}$$

$$= \frac{4}{\pi} \frac{B^2 m^4}{\rho \hbar^4} v^3. \tag{26}$$

The fact that Equation (5) yields the same result as Equation (3) means that the electron acts *as if* its available energy for interacting with the lattice had the coulomb value $e^2/(Kd)$.* Whether or not this has physical significance, it does point up the fact that the available energy for forming an energy well has the same numerical value for interactions both by deformation potential and by macroscopic coulomb fields.

If one should read physical significance into the use of Equation (5) for deformation-potential coupling, he would be led to conclude that β, by its definition, cannot exceed unity. This would, by Equation (25), put an upper limit on the value of the deformation potential B, particularly at the highest values of ω. For example, as shown in Part I,† a value of 10 eV for B already leads to values of β in excess of unity at $\omega \approx 10^{14}$/sec.

9. Nonpolar Optical Phonons

The rate of energy loss to nonpolar optical phonons is computed from Equation (3) in the same way as for acoustic phonons coupled by a deformation potential. The energy well follows from equating the electrical energy gained by the electron in deforming the lattice

† See *RCA Review*, March 1966, pp. 98-139.

* While the electronic charge does not appear explicity in Equation (26), it is contained implicitly in the magnitude of the deformation potential B.

to the elastic work done on the lattice. The deformation potential D for nonpolar optical phonons is defined as the shift in energy of the band edge per centimeter relative displacement of the two sublattices. (For acoustic phonons, B was defined in terms of the energy shift per unit strain.) Hence

$$\frac{1}{2} D\Delta x \approx \frac{1}{2} C(\Delta x)^2 \left(\frac{4}{3}\pi r^3\right) \tag{27}$$

and

$$\text{energy well} = \frac{1}{2} D\Delta x = \frac{3}{8\pi} \frac{D^2}{Cr^3}, \tag{28}$$

where Δx is the relative displacement of the sublattices, C is an elastic constant (per unit displacement), and r is the uncertainty radius.

Insertion of Equation (28) into Equation (3) yields

$$\frac{dE}{dt} = \frac{3}{4\pi} \frac{D^2 \omega^2}{C r^2 v}. \tag{29}$$

Use of the relations

$$r \approx \frac{\hbar}{mv}$$

and

$$C = \rho\omega^2 \tag{30}$$

converts Equation (29) to

$$\frac{dE}{dt} = \frac{3}{4\pi} \frac{D^2 m^2}{\rho \hbar^2} v. \tag{31}$$

Equation (31) is to be compared with Conwell's[5] result for $mv^2/2 \gg \hbar\omega$ and an isotropic effective mass,

$$\frac{dE}{dt} = \frac{1}{2\pi} \frac{D^2 m^2}{\rho \hbar^2} v. \tag{32}$$

The complete form of Conwell's published result is

$$\frac{dE}{dt} = \frac{m_t m_l^{1/2} D^2}{2^{1/2}\pi\hbar^2\rho}\left[(\bar{n}+1)\left(\frac{1}{2}mv^2-\hbar\omega\right)^{1/2} - \bar{n}\left(\frac{1}{2}mv^2+\hbar\omega\right)^{1/2}\right] \tag{33}$$

where $\bar{n} = (\exp\{\hbar\omega/kT\} - 1)^{-1}$ = density of thermal phonons.

Equation (31) can also be obtained, as was done in the preceding section for acoustic phonons, from Equation (5) using the value of β obtained in Part I.* This formulation is shown in Table II. The implication, as before, is that the energy available for interacting with the lattice is the coulomb energy $e^2/(Kd)$.

10. Intervalley Scattering

In the case of a many-valleyed band structure, an electron can lose energy to the lattice by making a transition from one valley to another (Herring[3]). For valleys widely separated in momentum space, the emitted phonons have a narrow spread in momentum around some high value of momentum. The phonons may be either acoustic or optical. An appropriate deformation potential is defined for each, just as in the preceding two sections, which is a measure of the relative energy shift of the two valleys per unit strain (per unit displacement of the two sublattices).

For energy loss to optical phonons the argument is the same as that used in the preceding section on nonpolar optical phonons and leads to the same result, namely,

$$\frac{dE}{dt} \approx \frac{3}{4\pi}\frac{D^2 m^2}{\rho\hbar^2} v. \tag{34}$$

For energy loss to acoustic phonons, the argument follows that already used for intravalley scattering and leads to Equation (20):

$$\frac{dE}{dt} = \frac{3}{4\pi}\frac{B^2}{Cr^2}\frac{\omega^2}{v}. \tag{35}$$

In the present case, however, ω has a large and relatively constant value obtained by matching $\hbar\omega/v_s$ with the change in crystal momentum suffered by the electron in making the transition between two valleys. This is in contrast to the argument in Section 8, where $\hbar\omega/v_s$ was equated to the local momentum at the electron in its valley. Hence we use only the two relations

* See *RCA Review*, March 1966, pp. 98-139.

$$r \approx \frac{\hbar}{mv}$$

and

$$C = \rho\, v_s^2$$

to obtain

$$\frac{dE}{dt} = \frac{3}{4\pi} \frac{B^2 m^2 \omega^2}{\rho\, v_s^2\, \hbar^2} v. \tag{36}$$

Equation (36) has the same dependence on v as Equation (34). Moreover, for large values of ω the coefficients of the two expressions have the same order of magnitude insofar as

$$\frac{B\omega}{v_s} = \frac{2\pi B}{\lambda} \approx D. \tag{37}$$

11. Plasmons and X-ray Levels

The treatment of energy loss to plasmons and to the deeper-lying atomic levels (x-ray levels) follows closely that of energy loss to polar optical phonons. In all three cases ω is substantially a constant, independent of wavelength, so that the problem approximates that of a set of independent oscillators. Hence, in this limit, they can be treated (for energy-loss calculations) either as independent oscillators (Bohr) or formally as a set of collective modes of oscillation (Bohm and Pines). Furthermore, since the condition for Equation (3) is $\omega\tau < 1$, the incident electron makes its transit in times short compared with the reaction time ω^{-1} of the medium. Thus, even the local oscillators "look" like free particles.

We begin with Equation (15) which was derived from Equation (3):

$$\frac{dE}{dt} = \frac{K_1 - K_2}{K_1 K_2} \frac{e^2\omega^2}{v} \ln \frac{r_o}{r_i}. \tag{38}$$

Here the real part of the dielectric constant that the electron sees is K_1 when its energy is less than $\hbar\omega$ and K_2 when its energy exceeds $\hbar\omega$.

For plasmons, K_1 is the usual electronic part of the dielectric constant and K_2 is close to unity. The $\hbar\omega$ of the plasmon appropriate to a filled valence band is taken to be of the order of 10 eV. Hence

$$\frac{K_1 - K_2}{K_1 K_2} \doteq \frac{K_1 - 1}{K_1} \doteq 1, \tag{39}$$

and Equation (38) becomes

$$\frac{dE}{dt} = \frac{e^2 \omega^2}{v} \ln \frac{r_o}{r_i}. \tag{40}$$

The outer radius, as before, is given by $2r_o = v\omega^{-1}$ and is that distance beyond which the medium is reversibly polarized and depolarized with negligible energy loss. The inner radius is normally taken to be a Debye screening length $v_F \omega^{-1}$ where v_F is the Fermi velocity. For shorter distances, the incident electron excites individual electrons rather than collective oscillations. Since, however, the individual excitations follow the same form as Equation (40) (see below), we can take r_i to be the uncertainty radius in order to write the total energy loss to plasmons and electron–electron collisions as

$$\frac{dE}{dt} = \frac{e^2 \omega^2}{v} \ln \frac{mv^2}{\hbar \omega}. \tag{41}$$

Equation (41) can also be obtained by an elementary treatment of particle-to-particle interactions. The momentum imparted to an electron at a distance r from the path of the incident electron is approximately

$$\begin{aligned} \Delta p &= \frac{e^2}{r^2} \Delta t \\ &\doteq \frac{e^2}{r^2} \frac{2r}{v} \\ &= \frac{2e^2}{rv}. \end{aligned} \tag{42}$$

The energy imparted is

$$\frac{(\Delta p)^2}{2m} = \frac{2e^4}{mv^2 r^2}. \tag{43}$$

The total energy imparted per unit time to a density of n electrons/cm^3 is

$$\frac{dE}{dt} = \frac{2ne^4}{mv^2} \int_{r_i}^{r_0} \frac{1}{r^2} (2\pi r)\, dr$$

$$= \frac{4\pi ne^4}{mv} \int_{r_i}^{r_0} \frac{dr}{r}$$

$$= \frac{e^2\omega_p{}^2}{v} \int_{r_i}^{r_0} \frac{dr}{r} \tag{44}$$

where $\omega_p{}^2 \equiv 4\pi ne^2/m =$ (plasma frequency)^2. The inner radius has the uncertainty value $\hbar/(mv)$. The outer radius is the adiabatic radius $v\tau$ where τ is the time for the electron gas to relax an electric field. By elementary considerations, $\tau = \omega_p{}^{-1}$. Hence Equation (44) becomes

$$\frac{dE}{dt} = \frac{e^2\omega_p{}^2}{v} \ln \frac{mv^2}{\hbar\omega_p}. \tag{45}$$

It is worth noting that the loss to polar optical phonons can be treated in the same way, namely as a particle-to-particle interaction between the incident electron and a gas of free ions. The result is a close match to Equation (11), since ω_p (ions) $\approx \omega_{optical\ phonon}$.

In the case of energy loss to the deeper-lying atomic levels, we make use of the following relation for the lower frequency dielectric constant K_1:

$$K_1 = 1 + \frac{\omega_p{}^2}{\omega_e{}^2}, \tag{46}$$

where $\omega_p{}^2 = 4\pi ne^2/m$ and $\hbar\omega_e$ is the excitation energy for the deep level and is larger than $\hbar\omega_p$. K_2 as before is near unity. Hence,

$$\frac{K_1 - K_2}{K_1 K_2} \doteq \frac{\omega_p{}^2}{\omega_e{}^2}, \tag{47}$$

and Equation (15) for these deeper levels becomes

$$\frac{dE}{dt} = \left[\frac{\omega_p{}^2}{\omega_e{}^2} \right] \frac{e^2\omega_e{}^2}{v} \ln \frac{mv^2}{\hbar\omega_e}$$

$$= \frac{e^2\omega_p{}^2}{v} \ln \frac{mv^2}{\hbar\omega_e}. \tag{48}$$

The factor ω_p^2/ω_e^2 is the fraction of the coulomb energy of the incident electron that is used in forming an energy well by polarizing the electrons in the deep-lying levels.

12. Impact Ionization

In the case of the deeper-lying atomic levels, the ground state energy was well defined and common to all of the electrons. The electrons in the valence band of a semiconductor are, by contrast, spread over a range of energies. Thus, as the energy of an impinging electron is increased, it is at first able to ionize only the electrons in the top edge of the valence band. Further increase in energy allows the impinging electron to reach deeper into the valence band. A first-order estimate of the rate of energy loss by the impinging electron can be obtained using the concepts already applied to polar optical phonons, plasmons, and the deeper atomic levels. In brief, we compute the real part of the dielectric constant as a function of the energy of the incident electron and use the relation

$$\frac{dE}{dt} = \left(\frac{1}{K_1} - \frac{1}{K_0} \right) \frac{e^2\omega^2}{v} \tag{49}$$

to compute the rate of energy loss. This is Equation (15) with the logarithmic factor taken to be unity since $mv^2/2 \approx \hbar\omega$. K_0 is the dielectric constant when the incident electron is just at the threshold of impact ionization, namely when its energy is equal to the energy E_g of the forbidden gap. K_1 is the dielectric constant for a somewhat higher incident energy $E = E_g + \Delta E$. We use the approximation that K_1 is proportional to the number of electrons in the valence band that are *too deep* to be ionized. Hence

$$\frac{K_1}{K_0} = \frac{n_0 - A(\Delta E)^{3/2}}{n_0}$$

and

$$\frac{K_0 - K_1}{K_0} = \frac{A(\Delta E)^{3/2}}{n_0}$$

$$= \left(\frac{\Delta E}{E_0} \right)^{3/2}, \tag{50}$$

where n_0 = total number of electrons in the valence band and E_0 is an energy comparable with the width of the valence band and defined by $n_0 = AE_0{}^{3/2}$. We take $\hbar\omega \approx E = E_g + \Delta E$ and $v = (2E/m)^{1/2}$. With these relations Equation (49) becomes

$$\frac{dE}{dt} = \frac{e^2 m^{1/2}}{2^{1/2}\hbar^2 K_1} \left[\frac{(\Delta E)(E_g + \Delta E)}{E_0} \right]^{3/2}. \tag{51}$$

For $\Delta E \ll E_g$, Equation (51) matches the $(\Delta E)^{3/2}$ dependence obtained by Motizuki and Sparks.[6] It is worth noting also that if we were to let E_g become small in comparison with ΔE in order to approach the conditions of a metal, the dependence on ΔE becomes $(\Delta E)^3$, which is that found by Quinn[23] for metals.

13. Cerenkov Radiation

When the velocity of a high-speed electron exceeds the velocity of light in a solid, a new channel for energy loss, namely electromagnetic radiation, is opened up. There are various conceptual approaches to the analysis of Cerenkov radiation. Quantum mechanically, the radiation may be looked upon as induced by the zero point fluctuations in the radiation field. The radiation may also be computed classically by resolving the pulse of electric field, as seen by an element of the medium when the electron passes nearby, into its Fourier components.

What we wish to show here is that the expression for the total Cerenkov radiation matches Equation (5) for velocities close to the velocity of light in vacuum. The expression for the total Cerenkov radiation (see e.g. Schiff[13]) is

$$\frac{dE}{dt} = \int_0^\omega \frac{e^2\omega^2 v}{c^2} \left(1 - \frac{c^2}{\bar{n}^2 v} \right) d\omega. \tag{52}$$

Here v is the velocity of the electron, c is the velocity of light in vacuum, $\bar{n}$ is the index of refraction and is a function of ω. The upper limit of integration is the highest frequency for which the index of refraction exceeds unity. In the limit for $v \to c$, and using the approximation of a constant index n, Equation (52) becomes

[23] J. J. Quinn, "Range of Excited Electrons in Metals," *Phys. Rev.*, Vol. 126, p. 1453, 15 May 1962.

$$\frac{dE}{dt} = \left(\frac{\bar{n}^2 - 1}{\bar{n}^2}\right) \frac{e^2\omega^2}{2c}$$

$$= \left(\frac{\epsilon_\infty - 1}{2\epsilon_\infty}\right) \frac{e^2\omega^2}{c}. \tag{53}$$

Equation (53) has the same form as Equation (5) with $\beta = (\epsilon_\infty - 1)/\epsilon_\infty$.

14. Cerenkov Angle

According to Equation (53), the spatial rate of loss of energy by the high-velocity electron is of the order of $e^2\omega^2/c^2$ or less than 0.1 eV per wavelength. Since the dominant emission is in the visible or near ultraviolet, the photons have energies of several electron volts. Hence, the energy from a number of wavelengths must combine coherently to form a single emitted photon. From geometric considerations this can occur only along a well-defined angle θ measured from the line of flight such that

$$\cos\theta = \frac{c}{\bar{n}v}. \tag{54}$$

We wish to point out that the Cerenkov angle also follows readily from the elementary relations (Equations (6) and (7) of Chap. III) for power transfer. From these relations, the fraction of the total radiated power that is associated with forward momentum is v_s/v where v_s is the velocity of wave propagation.

Consider the total radiated power being emitted along a cone of angle θ surrounding the line of flight of the electron. It will then be true that

$\cos\theta$ = fraction of radiated power associated with forward momentum,

$\sin\theta$ = fraction of radiated power associated, by symmetry, with zero momentum.

Hence, the cone angle satisfies the Cerenkov condition,

$$\cos\theta = \frac{v_s}{v} = \frac{c}{\bar{n}v}.$$

15. Heuristic Argument for Relating Acoustoelectric Effects and Energy Loss by Fast Carriers

In the following argument we show how the classical acoustoelectric expressions for rates of amplification or attenuation of coherent acoustic waves, discussed in Chap. III, need to be modified to arrive at the quantum mechanical form for the rates of emission of phonons by individual electrons. In so doing, some clarification is provided for the fact that the parameter β, the ratio of electrical to total energy, is common to both sets of phenomena.

We begin with Equation (10) of Chap. III (including the diffusion term) for the acoustoelectric effect;

$$\frac{dE}{dt} = \beta E_t \omega \frac{\omega_r/\omega_c}{(1+\omega_D/\omega_c)^2 + (\omega_r/\omega_c)^2}, \tag{55}$$

and make use of the following relations:

(a)
$$\omega_c^{-1} = \frac{K}{4\pi n e \mu} = \frac{Km}{4\pi n e^2 \tau_c}$$

where τ_c = collision time and n = density of free carriers.

(b) $$\omega_r = \frac{v - v_s}{v_s}\omega = \frac{v}{v_s}\omega \quad \text{for } v \gg v_s.$$

(c) The energy-loss expressions of Table II hold for $T = 0$. Hence we can take

$$\omega_D = \frac{4\pi^2 kT\mu}{\lambda^2 e} = 0$$

(d) Since we are dealing with the rates of loss of individual electrons in Table II we can take Equation (55) in the low-density limit where $\omega_r/\omega_c \gg 1$.

With these substitutions Equation (55) becomes

$$\frac{dE}{dt} = 4\pi\beta E_t \frac{ne^2\tau_c}{Km}\frac{v_s}{v}, \tag{56}$$

and the energy loss per electron is

$$\frac{dE_1}{dt} \equiv \frac{1}{n}\frac{dE}{dt} = 4\pi\beta E_t \frac{e^2\tau_c}{Km}\frac{v_s}{v}. \tag{57}$$

Equation (57) is a classically derived expression for which

$$\text{mean free path of electron } l \ll \frac{\lambda}{2\pi}. \tag{58}$$

On the other hand, the perturbation treatment of the emission of phonons by individual electrons is valid for $l > \lambda/(2\pi)$. If the two have a common region, it should be near $l = \lambda/(2\pi)$. Hence we evaluate Equation (57) at the junction:

$$l = \frac{\lambda}{2\pi}$$

or

$$\tau_c \equiv \frac{l}{v} = \frac{\lambda}{2\pi v}. \tag{59}$$

Equation (59) inserted in Equation (57) gives

$$\frac{dE_1}{dt} = 2\beta E_t \frac{e^2\lambda}{Km}\frac{v_s}{v^2}. \tag{60}$$

E_t is normally the total energy density of a coherent sound wave. In the present argument the only available energy is that of the incoherent zero-point vibrations. Their energy density in the interval $\lambda \pm \lambda/2$ and resolved along the axis of electron motion is

$$E_t = \frac{1}{3}\frac{8\pi\,\hbar\omega}{\lambda^3}. \tag{61}$$

With this value for E_t, Equation (60) becomes

$$\frac{dE_1}{dt}=\frac{16\pi}{3}\,\frac{\beta e^2\,\hbar\omega v_s}{Km\lambda^2 v^2}. \tag{62}$$

We use the momentum condition

$$2mv=\frac{\hbar\omega}{v_s}$$

and the definition

$$\lambda=\frac{2\pi v_s}{\omega}$$

to convert Equation (62) to

$$\frac{dE_1}{dt}=\frac{8}{3\pi}\,\beta\,\frac{e^2\omega^2}{Kv}. \tag{63}$$

This is the essential form of all of the expressions for rates of loss of energy in Table II.

16. General Remarks

A. The Role of Quantum Constraints

The core of the argument for energy loss, Equations (3) and (5), is classical. Quantum effects are introduced as constraints.

For example, the major contribution of quantum mechanics in the case of polar optical phonons, electronic excitations, and plasmons is to determine an energy threshold $mv^2/2 > \hbar\omega$ at which significant energy loss sets in. A secondary contribution in each case is to determine a momentum-controlled inner radius, as opposed to the classical energy-controlled inner radius, in the logarithmic factor.

In the case of piezoelectric phonons, the maximum value of ω is constrained quantum mechanically to a value satisfying the momentum relation

$$2mv=\frac{\hbar\omega}{v_s}.$$

The energy relation is less restrictive,

$$\frac{mv^2}{2}=\hbar\omega.$$

The momentum relationship reduces the highest phonon frequency ω with which an electron can interact to values of the order of v_s/v times smaller than that allowed energetically. The same is true for the acoustic phonons coupled by deformation potentials.

In the case of nonpolar optical phonons, the quantum constraints impose both an energy threshold, as for polar optical phonons, and an upper limit to the wave vector of the dominant phonon interaction, as for acoustic phonons.

Note that in the classical model the incident electron moves in a straight line emitting energy symmetrically and simultaneously around its line of flight. It is, by definition, not scattered. Scattering is introduced by imposing the quantum constraint that this same average rate of emission of energy takes place via the emission of a *sequence* of *individual* phonons.

B. *Summary of Rates of Energy Loss*

Figure 3 summarizes the rates of loss of energy to the various types of phonons as a function of electron energy. Representative values were chosen for the bracketed factors in Table II. The factors are the ratios of electrical to total energy for the several phonons.

To represent energy loss to polar optical phonons in the alkali halides the following values were used:

$$\frac{\epsilon_0 - \epsilon_\infty}{\epsilon_0} = 0.5,$$

$$\epsilon_\infty = 2.5,$$

$$\omega = 5 \times 10^{13}/\text{sec},$$

$$\frac{m}{m_0} = 1.$$

At the other extreme, the following values were used to approximate energy loss to polar optical phonons in a material such as InSb:

$$\frac{\epsilon_0 - \epsilon_\infty}{\epsilon_0} = 0.05,$$

$$\epsilon_\infty = 20,$$

$$\omega = 5 \times 10^{13}/\text{sec},$$

$$\frac{m}{m_0} = 10^{-2}.$$

For piezoelectric phonons a value of $\epsilon_p{}^2/KC$ of 0.05 was used to approximate CdS.

For deformation-potential acoustic phonons, B was chosen to be 10 eV or 1.6×10^{-1} erg. This value is of the order of that found in germanium. For nonpolar optical phonons, D was taken to be 4×10^8 eV/cm. This value is quoted by Conwell for germanium.

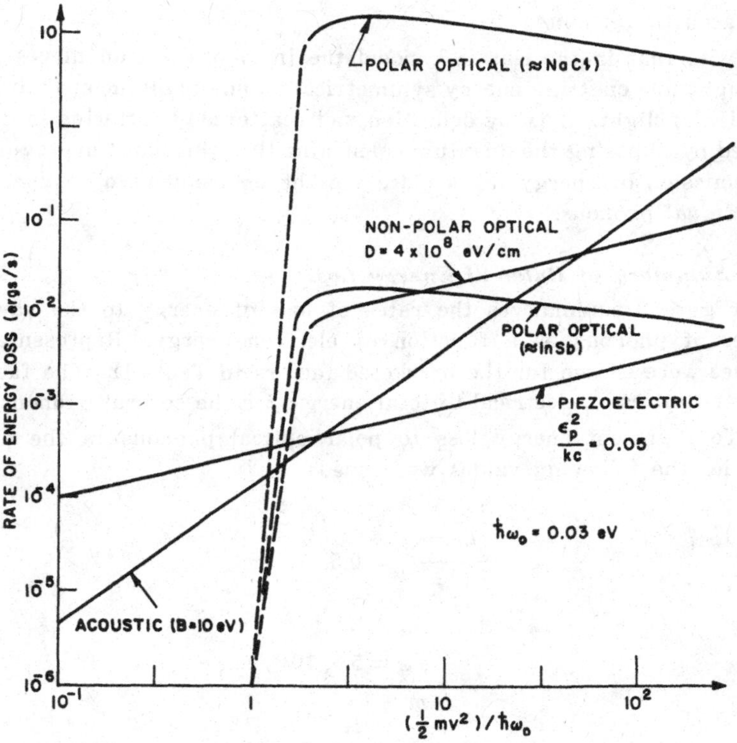

Fig. 3—Representative rates of energy loss to various modes of electron-phonon coupling.

The values chosen for the other significant parameters are

$$K = 10,$$

$$v_s = 3 \times 10^5 \text{ cm/sec},$$

$$\rho = 10 \text{ gms/cm}^3,$$

$$\frac{m}{m_o} = 0.2,$$

$$\omega_{\text{optical}} = 5 \times 10^{13}/\text{sec}.$$

The energy scale for the emitting electron is plotted in units of $\hbar\omega_{optical}$ where $\hbar\omega_{optical} = 0.03$ eV. The width of the conduction band is taken to be 10 eV.

For comparison, the order of magnitude of the rate of loss to Cevenkov radiation is 10^3 ergs/sec and to plasmons is 10^5 ergs/sec.

One virtue of expressing the rates of energy loss in the format of Table II is that the upper limit to the rate of energy loss to any type of radiation is clearly defined by

$$\frac{dE}{dt} = \frac{e^2\omega^2}{v} \ln \frac{2mv^2}{\hbar\omega}. \tag{64}$$

This is the maximum rate of energy loss obtained by setting $\beta = 1$ and $K = 1$. In effect, all of the vacuum coulomb energy of the electron is used to form an energy well.

The entries in Table II can be used to compute the mean free path for emitting a quantum of radiation, since

$$l \equiv \frac{\hbar\omega}{dE/dx} = \frac{\hbar\omega v}{dE/dt}$$

$$\approx K \frac{r_H}{\beta} \frac{mv^2}{\hbar\omega} \frac{m_o}{m} \times \begin{cases} 1 \\ \text{or} \\ 1/[\ln (2mv^2/\hbar\omega)] \end{cases} \tag{65}$$

where r_H is the Bohr radius of the hydrogen atoms and m_0 is the mass of a free electron.

Insertion of the values of β, taken from Table II, into Equation (65) yields the following dependencies of the mean free path on the velocity of the electron when the thermal density of phonons is negligible:

polar optical	$l \propto \dfrac{v^2}{\ln (2mv^2/\hbar\omega)}$
non-polar optical	$l \propto$ constant
acoustic (piezoelectric)	$l \propto v$
acoustic (deformation potential)	$l \propto v^{-1}$

ACKNOWLEDGMENTS

The concepts outlined in this paper have been developed over a period of several years. During this time I have had the generous guidance and criticism of many of my colleagues. By far, the major guidance has come from Professor George Whitfield of Pennsylvania State University to whom I am deeply indebted for the patience and insight needed to bridge the gap between my elementary classical concepts on the one hand and the formal rigor of perturbation theory on the other.

This work was supported by the U. S. Army Research Office—Durham, Durham, North Carolina under Contract No. DA-31-124-ARO(D)-84 and by the U. S. Army Engineer Research and Development Laboratory, Fort Belvoir, Virginia, under Contract No. DA44-009-AMC-1276(T); and by RCA Laboratories, Princeton, N. J.

Reprinted from *RCA Review*, 30 (1969) 435,

Chapter V

Field and temperature dependence of electronic transport

PREAMBLE

THE MAJOR purpose of this chapter is to trace the events that lead at high fields to hot electrons, impact ionization, and dielectric breakdown; and to trace the events in terms as simple and as physically graphic as possible. Certain concepts have emerged that were not evident at the outset, at least to this writer, and will be outlined here.

The rates of exchange of energy between electrons and a medium are characterized by a coupling constant β, which was previously defined as the fraction of the available coulomb field energy of the electron that can be transferred to the medium. It was shown that in the case of energy exchanges between electrons on the one hand and phonons or acoustic waves on the other, β is equal to the ratio of electrical to total energy of the corresponding acoustic wave. Further, when a fast electron excites resonances such as polar optical phonons, plasmons, or x-ray levels, β can also be written as $(K_L - K_H)/K_L$, where K_H and K_L are, respectively, the dielectric constants on the high and low sides of the resonance.

The coupling constant β was found to be valid for the classical acoustoelectric effect in which macroscopic acoustic waves are amplified by a drifting stream of electrons as well as for the quantum processes in which hot electrons emit phonons to a phonon-free or zero-temperature medium. The acoustoelectric effect is a process of *induced* emission, whereby the rate of loss of energy to the acoustic wave is proportional to the energy of the acoustic wave or, in quantum language, to the number of phonons in the wave. On the other hand, the emission of phonons into a zero-temperature lattice by an energetic electron is conventionally called *spontaneous* emission.

The term spontaneous was Bohr's interpretation of Einstein's phrase "emission without excitation by external causes"† of a photon by an excited atom in a field-free surround. In quantum terminology, the rate of emission by an excited atom is proportional to $n_{Ph} + 1$ where n_{Ph} is the occupation number for photons of the emitted frequency and represents the induced part of the emission. It was conceptually reasonable from classical physics that the presence of real photons would indeed provoke or induce the atom to emit as well as to absorb. The term unity in $n_{Ph} + 1$, however, meant that even at zero temperature, in the absence of real photons in the field, the atom would still emit—apparently spontaneously, because there was nothing in the field to provoke it to emit.

It is for this reason that spontaneous emission has come to have a certain mystery attached to it. It is not uncommon to find some writers,[1] apparently in an attempt to avoid a logical vacuum, attributing spontaneous emission to the presence of zero-point vibrations. This association has two unfortunate consequences. First, it raises the question[1] of why zero-point vibrations, whose energy content is only a

† See, e.g., *The Conceptual Development of Quantum Mechanics*, by M. Jammer, McGraw-Hill Book Co., N. Y. (1966).

[1] L. I. Schiff, *Quantum Mechanics*, McGraw-Hill Book Co., Inc., New York, p. 388, 1949.

half quantum, should be equally as effective as real photons whose energy content is a whole quantum. Even more, it suggests that spontaneous emission is a purely quantum mechanical phenomenon that cannot be understood classically because zero-point vibrations have no classical counterpart.

In contrast to certain, at least historical, ambiguities attached to the concept of spontaneous emission in vacuum, the concept of spontaneous emission in a solid medium as described in Chap. III has a simple, classical foundation. An electron in a solid polarizes or distorts the solid so that energy is transferred from the coulomb field of the electron to the surrounding solid medium. When the electron moves sufficiently fast, it leaves behind a trail of polarization whose energy content is equal to the average rate of emitting phonons (or other excitations). Spontaneous emission is no more mysterious than induced emission. In fact, either one implies the other. The fact that the medium can act on the electron, via the electric fields of its phonons, implies that the electron must be able to act upon (polarize) the medium and emit phonons. The electron and the medium are two systems. If we postulate action of one system (medium) on the other (electron), we are committed to the action of the second system (electron) on the first (medium). In brief, the interaction between electron and medium is a symmetric, two-sided connection that should be describable by a single coupling constant. The coupling constant β, which we have already outlined, does indeed characterize both spontaneous and induced emission of phonons by electrons in solids.

The major content of the present chapter is the computation of the mean energy of electrons under the application of fields sufficient to raise that mean energy above the kT_L of the lattice.* These are commonly called hot electrons. The mean energy can be characterized by an equivalent temperature T_e by equating the mean energy to kT_e. However, this assignment does not ensure that the distribution of energies around the mean value satisfies the Boltzmann form $\exp(-E/kT_e)$. Such is not, in general, the case, and some estimates based on simplifying assumptions will be made to obtain a measure of the actual distributions.

Prior to the analysis of hot electrons, we outline a general method for obtaining the mobilities of electrons in thermal equilibrium with the lattice. The method represents a "back-door" approach to mobility. The usual approach computes by perturbation theory the scattering of

---

* For a treatment of hot electrons by perturbation theory, see *High Field Transport in Semiconductors*, by E. M. Conwell, Solid State Phys. Suppl., Academic Press, N. Y. (1967).

electrons by real phonons, that is, the *induced* emission and absorption of phonons. The present approach begins, instead, with the *spontaneous* emission of phonons and relates this to the induced emission via detailed balancing that states that the total electron–phonon interaction is $2n_{Ph}+1$ times the spontaneous interaction (here, n_{Ph} is the occupancy for phonons). We use this back-door approach for two reasons. First, we have already carried out a relatively simple semiclassical formalism in Chap. II for computing the spontaneous emission. The formalism is couched in real space. I do not know of a comparable formalism for induced emission. The second reason is that the back-door approach is a back door only by convention. As we have already pointed out in this preamble, induced and spontaneous emission are two aspects of the *same* physical interaction between electrons and phonons. We are free to choose either aspect as a means for computing the total interaction.

Temperature Dependence of Mobility in Semiconductors

Let dE/dt be the average rate at which an electron of given energy spontaneously emits energy to a given class of phonons. The actual emission is via a series of phonons randomly spaced in time. Hence, the average time required to emit one phonon is

$$\tau_e = \frac{\hbar\omega}{dE/dt} \tag{1}$$

From detailed balancing, we know that the total interaction between electrons and phonons occurs at the rate of $2n_{Ph}+1$ times that for spontaneous emission. Here, n_{Ph} is the occupancy for phonons of frequency ω,

$$n_{Ph} \equiv \left[\exp\left(\frac{\hbar\omega}{kT}\right) - 1\right]^{-1}$$

$$\doteq \frac{kT}{\hbar\omega} \text{ for } kT \gg \hbar\omega. \tag{2}$$

The average collision time for the total interaction is, then,

$$\tau_c = \frac{1}{2n_{Ph}+1}\,\tau_e$$

$$\doteq \frac{\hbar\omega}{2kT}\,\tau_e \text{ for } kT \gg \hbar\omega$$

$$= \frac{(\hbar\omega)^2/(2kT)}{(dE/dt)} \tag{3}$$

Now, the spontaneous emission (see Table I) is a strongly increasing function of ω, so that Equation (3) is dominated by the highest-frequency phonons. Each such collision is sufficient to randomize the momentum of the electron and to constitute one collision time for the purpose of computing mobility. We can, accordingly, use Equation (3) to write the mobility as

$$\begin{aligned} \mu &\equiv \tau_c \frac{e}{m} \\ &= \frac{e\,[\hbar\omega/(mv_t)]^2}{dE/dt} \end{aligned} \tag{4}$$

where v_t is the mean thermal velocity of the electron and dE/dt is the spontaneous rate of energy loss by an electron having thermal energy.

We take the v_t dependence of dE/dt from Table I and compute the following temperature dependencies of mobility from Equation (4) for optical phonons for which $\omega =$ constant.

Polar Optical:

$$\mu \propto v_t^{-1} \propto T^{-1/2}, \text{ for } kT > \hbar\omega \tag{4a}$$

Non-Polar Optical:

$$\mu \propto v_t^{-3} \propto T^{-3/2}, \text{ for } kT > \hbar\omega \tag{4b}$$

For optical phonons and for $kT < \hbar\omega$, the dominant factor in the mobility is an exponential factor

$$\mu \propto \exp\left(-\frac{\hbar\omega}{kT}\right) \tag{5}$$

that is a measure of the fraction of electrons having enough energy

Table I—Time rates of energy loss dE/dt by electrons of velocity v. The middle column expresses these rates formally as $\beta(e^2\omega^2/Kv) \times$ geometric factor (derived in Chap. III), where β is the coupling constant and is set off by square brackets. The right-hand column gives the v-dependence of dE/dt after inserting the quantum constraints noted in the formal expressions. The numerical factors (½, ¾, π/4) preceding the square brackets are the ratio of the quantum mechanically derived results in the literature to our semi-classically derived results.

Phonon and Coupling	Formal Expression for dE/dt	Final Expression for dE/dt in the Form Av^n
Polar Optical (Polarization Field)	$\left[\frac{\epsilon_o - \epsilon_\infty}{\epsilon_o}\right] \frac{e^2\,\omega^2}{\epsilon_\infty v} \ln\left(\frac{2mv^2}{\hbar\omega}\right)$	$\sim v^{-1}$
Non-Polar Optical (Deformation Potential)	$\frac{1}{2}\left[\frac{\pi K D^2}{\rho e^2\,\omega^2\lambda^2}\right] \frac{e^2\omega^2}{Kv}$ Note: $\frac{\lambda}{2\pi} = \frac{\hbar}{2mv}$	$\frac{D^2 m^2}{2\pi\rho\hbar^2}\, v$
Acoustic (Deformation Potential)	$\frac{3}{4}\left[\frac{B^2\omega^2 K}{4\pi e^2 \rho v_s^4}\right] \frac{e^2\omega^2}{Kv}$ Note: $\frac{\hbar\omega}{v_s} = 2mv$	$\frac{B^2 m^4}{\pi\rho\hbar^4}\, v^3$
Acoustic (piezoelectric)	$\frac{\pi}{4}\left[\frac{\epsilon_p^2}{KC}\right] \frac{e^2\omega^2}{Kv}$ Note: $\frac{\hbar\omega}{v_s} = 2mv$	$\frac{\pi\epsilon_p^2\, e^2 m^2 v_s^2}{K^2 C\hbar^2}\, v$

Definitions for Table I

ϵ_o = low-frequency dielectric constant

ϵ_∞ = high-frequency (optical) dielectric constant

K = electronic part of dielectric constant

ϵ_p = piezoelectric constant

C = elastic modules (dynes/cm^2)

ρ = density (grams/cm^3)

v_s = phase velocity of sound

v = velocity of electrons

ω = angular frequency of phonon

B = deformation potential (electron volts in ergs/unit strain)

D = optical deformation potential (electron volts in ergs per centimeter relative shift of sublattices)

m = effective mass of electrons

($\hbar\omega$) to emit an optical phonon. Alternatively, the exponential factor is a measure of the occupancy of optical phonons that can be absorbed by the electrons.

In the case of acoustic phonons, the highest phonon frequency to be inserted in Equation (4) is given by the conservation of crystal momentum,

$$\frac{\hbar\omega}{v_s} \approx 2mv_t, \tag{6}$$

where v_s is the phase velocity of sound. Equation (6) inserted into Equation (4) yields

$$\mu \approx \frac{4ev_s^2}{dE/dt}. \tag{7}$$

We use Table I and Equation (7) to obtain
Acoustic (Piezoelectric Coupling)

$$\mu \propto v_t^{-1} \propto T^{-1/2}. \tag{8a}$$

Acoustic (Deformation Potential Coupling)

$$\mu \propto v_t^{-3} \propto T^{-3/2}. \tag{8b}$$

These mobilities and their temperature dependencies are based on a constant, isotropic effective mass for the electron. An anisotropic effective mass introduces some averaging process for the effective mass but does not alter the temperature dependence. An energy-dependent effective mass affects the temperature dependence of mobility in more or less obvious ways. That is, an effective mass that increases with increasing energy will cause the mobility to decrease more rapidly with increasing temperature.

Mobility in Metals

We treat the problem of mobility in metals separately from that in semiconductors for several reasons. The disparity between the mathematical level of a formal treatment for metals in terms of perturbation theory[2] and of a real-space model[3] is even more pronounced than for semiconductors; the dielectric constant takes on a different form for metals; the temperature dependence is a more involved problem; and finally the real-space model gives an easy insight into the Fröhlich mechanism for superconductivity.

High Temperatures

We begin with temperatures greater than the Debye temperature so that Equation (3) gives the general form for the mobility collision time.

$$\tau_c = \frac{(\hbar\omega)^2/(2kT)}{dE/dt}. \tag{9}$$

Similarly dE/dt has the general form (see Table I)

$$\frac{dE}{dt} \approx \beta \frac{e^2\,\omega^2}{Kv_F}, \tag{10}$$

where v_F is the velocity of an electron at the Fermi surface. Combination of Equations (9) and (10) then yields the mobility

$$\mu \equiv \tau_c \frac{e}{m} = \frac{K\hbar^2 v_F}{2\beta emkT}. \tag{11}$$

In the case of semiconductors or ionic solids in particular, K is the electronic part of the dielectric constant. It enters into Equation (10) because we are interested in the force of the incident electron on the surrounding ions. This force is due to the vacuum coulomb field of the electron reduced by the electronic part of the dielectric constant. Similarly, in a metal K represents the shielding action of the ensemble of metal electrons on the force between any one electron and the surrounding ions. For distances greater than a Debye length this shielding can be written

---

[2] F. Bloch, "Über die Quantenmechanik der Elektronen in Kristallgittern," *Zeitschrift fur Physik*, Vol. 52, p. 555, 1929; J. M. Ziman, *Electrons and Phonons*, Oxford Univ. Press, New York, p. 371, 1960.

[3] A. Rose and A. Rothwarf, Proc. IX Int. Conf. on Physics of Semicond., Moscow, Vol. 2, p. 694, 1968. Published by Nauka, Leningrad.

$$K = 1 + \left(\frac{L}{L_D}\right)^2 \doteq \left(\frac{L}{L_D}\right)^2. \tag{12}$$

We take for L the distance of closest approach between an electron and ion, and for this distance we take the uncertainty radius of the electron. Thus,

$$L \approx \frac{\hbar}{mv_F}, \tag{13}$$

and

$$L^2{}_D = \frac{E_f}{4\pi ne^2}, \tag{14}$$

where n is the density of electrons in the metal and related to E_F by

$$n = \frac{8^{1/2}}{3\pi^2} \frac{(mE_F)^{3/2}}{\hbar^3} \doteq \frac{1}{\pi^2} \cdot \frac{(mE_F)^{3/2}}{\hbar^3}. \tag{15}$$

Combination of Equations (12)-(15) yields

$$K = \frac{4e^2}{2^{1/2}\pi\hbar v_F} \doteq \frac{e^2}{\hbar v_F}. \tag{16}$$

Insertion of Equation (16) into Equation (11) then gives

$$\mu \doteq \frac{e\hbar}{2\beta mkT}. \tag{17}$$

The value of β is estimated to be about 0.5 and is based on the following argument. One definition of β was given in Chap. II as the ratio of electrical to total energy of the emitted phonons or of their equivalent macroscopic sound waves. For the jellium model of a metal, all of the energy of a sound wave is electrical, and β would be unity. But the jellium model of a metal is a plasma of ions immersed in a smoothed out sea of negative charge. A plasma has longitudinal vibrations but no transverse vibrations. The frequency of the latter is zero, since no work is required to shear a plasma. In an actual metal, the frequencies of transverse and longitudinal vibrations are comparable. The energy for the transverse vibrations comes from an "elastic" force or energy not present in the jellium model. The elastic energy must be comparable to the electrical energy in order that the frequencies of transverse vibrations (which involve only the elastic energy) be comparable to the longitudinal frequencies (which involve the sum of elastic and electrical energies). Hence, the ratio of electrical to total energy is of order 0.5, and this is the value of β.

Equation (17), with $\beta = 0.5$, becomes

$$\mu \doteq \frac{e\hbar}{mkT}. \qquad (18)$$

Note that the mobility collision time in Equation (18) is

$$\tau_c \doteq \frac{\hbar}{kT}. \qquad (19)$$

This is simply the smallest value of τ_c that would be permitted by the uncertainty relation

$$\Delta E \, \Delta t \geqq \hbar \qquad (20)$$

if we take $\Delta E = kT$. As suggestive as this interpretation is, it is not generally valid. According to Peierls[4], the ΔE of the uncertainty relation is of the order of E_F rather than kT. In any event, values of τ_c considerably smaller than that given by Equation (20) with $\Delta E = kT$ are encountered in polyvalent metals.

Equations (18) and (19) yield the experimental values of the monovalent metals within a factor of two. It is also true that the relatively complex expression for mobility derived by perturbation theory[2] reduces to Equations (18) and (19) in the jellium approximation for monovalent metals with $m = m_o$, the free electron mass.

Low Temperatures

The conductivity of metals at temperatures well below the Debye temperature has been observed and computed (Bloch theory) to vary

[4] R. E. Peierls, *Quantum Theory of Solids*, Clarendon Press, Oxford, England, p. 140, 1955.

as T^{-5}. We outline here the factors in the semiclassical argument that lead to the same temperature dependence.

At low temperatures, owing to the obvious energy constraints imposed by the Fermi distribution, the maximum phonon energy an electron can emit (or absorb) on the average is

$$\hbar\omega = kT$$

or

$$\lambda \propto T^{-1}. \tag{21}$$

From Equation (12), the screening factor will then vary as

$$K \propto \lambda^2 \propto T^{-2}, \tag{22}$$

since λ is now the distance of closest approach between the electron and the ions to which it can lose energy.

A further consequence of the low temperature is that the emission or absorption of a single phonon of momentum,

$$\hbar q = \frac{\hbar\omega}{v_s} \approx \frac{kT}{v_s}, \tag{23}$$

is no longer sufficient to completely scatter the momentum $\hbar k_F$ of a Fermi electron. The number of phonons required to turn the Fermi momentum through about 90° *if all of the phonons were added in the same sense* is, of course, k_F/q. Actually, the emission of phonons is a random process whose average contribution to the rotation of the Fermi momentum is zero. Hence, the rotation of the Fermi momentum must come from the rms deviation from the average. The rms deviation is proportional to the square root of the average number of phonons. Hence, $(k_F/q)^2$ phonons are needed on the average to randomize the momentum of a Fermi electron. The collision time of Equation (9) is, accordingly, increased by the factor

$$\left(\frac{k_F}{q}\right)^2 \propto \omega^{-2} \propto T^{-2} \tag{24}$$

The combined T dependencies of Equations (11), (22), and (24), then, lead to the appropriate T^{-5} at low temperatures.

Low-Temperature Pairing Energy

On the one hand, a Fermi electron at very low temperature cannot, by the exclusion principle, emit a high-energy phonon. On the other hand, it is clear from our semiclassical argument that the Fermi electron must still disturb the lattice as if it were in the act of emitting such a phonon. The "conflict" can be resolved if a second electron of equal and opposite momentum to that of the first electron cooperates with the first electron to absorb the latent energy in the trail of the first electron. This is the type of attractive energy first proposed by Fröhlich[5] as the source of the phenomenon of superconductivity.

The absorption by a second electron of phonon energy "virtually" emitted by a first electron is very much in the nature of a two-electron polaron. For example, the electron in the conventional one-electron polaron does not have sufficient energy to emit an optical phonon. Nevertheless, the electron disturbs the lattice as if it were about to emit such a phonon. The resolution here is that the electron re-absorbs its own phonons in a process known as virtual emission. The classical parallel is that of a slowly rolling ball on a rubber membrane. The leading edge of the ball does work on the membrane as if it were about to radiate vibrational energy to the membrane. At the same time, the trailing edge of the ball absorbs a comparable amount of energy from the previously stretched membrane. It is as if the trailing edge absorbs the energy emitted by the leading edge. While these two energies are only approximately equal in the macroscopic classical domain, they are identical in the quantum regime.

For the conventional one-electron polaron, the electron is moving slowly enough to re-absorb its own phonons. The Fermi electron in a metal, on the other hand, is moving too fast to absorb the phonons it "emits," and must make use of a second electron to form a "two-electron polaron."

We estimate the attractive energy for the two-electron problem by computing the depth of the energy trough left behind in the trail of the first electron.

The energy per unit *length* (see Equation (10)) in this trail is

$$\frac{dE}{dx} = \beta \frac{e^2 \omega^2}{K v_F^2}, \tag{25}$$

[5] H. Fröhlich, "Theory of the Superconducting State. I. The Ground State at the Absolute Zero of Temperature," *Phys. Rev.*, Vol. 79, p. 845, Sept. 1, 1950; *Rept. Prog. in Phys.*, Vol. 24, p. 1, 1961.

or, using Equation (16) for K,

$$\frac{dE}{dx} = \beta \frac{\hbar \omega^2}{v_F}.$$

The second electron then "samples" an element of length of the trough on the order of the uncertainty radius of the electron $\hbar/mv_F$. The attractive energy sensed by the second electron is then*

$$\begin{aligned} E &= \frac{\hbar}{mv_F} \frac{dE}{dx} \\ &= \beta \frac{\hbar^2 \omega^2}{mv_F^2} \\ &= \frac{\beta}{2} \left(\frac{\hbar \omega}{E_F} \right) \hbar \omega. \end{aligned} \tag{26}$$

For reasonable values of the parameters, the attractive energy is equivalent to a few degrees Kelvin. If, on the other hand, dE/dx is estimated from the experimental values of resistivities of polyvalent metals, the interaction energy of Equation (26) ranges up to about 10°K, namely, the range of transition temperatures for many superconductors.

General Comment on Mobility in Amorphous Materials

It is worth noting that the present semiclassical approach to mobility is particularly suited to estimating the transport properties of amorphous materials, in which there is a growing interest. The major factor in computing mobility is the rate of loss of energy, dE/dt, by an electron. The rate of loss of energy is largely insensitive to the periodicity of the medium, and should be equally valid for both crystalline and amorphous materials. The major uncertainty is in the choice of effective electron mass. To this extent one would expect that the *phonon contribution* to mobility would remain of the same order of magnitude in amorphous as in crystalline materials. This is born out in the case of metals, where the resistivity usually increases by only a

* A result of comparable magnitude was derived by K. S. Singwi by a more detailed argument, *Phys. Rev.*, Vol. 87, p. 1044 (1952).

factor of about two in going from the solid to the liquid state. Even this factor of two increase may be ascribed to the increase in elastic scattering due to the increased disorder. The temperature-dependent part of the resistivity, which is a measure of the phonon contribution, is substantially unchanged on melting.

It is also true (see Chap. III, p.) that the rate of loss of energy by an electron in an ionic crystal is substantially the same whether it is computed conventionally by perturbation theory (electron–phonon interaction) or by treating the ions as a dense plasma and computing the rate of energy loss to plasma excitations. Periodicity, in brief, plays only a secondary role.

Mean Energies and Mobilities of Hot Carriers

In general, when an electric field is applied to a semiconductor, the average energy of the free carriers must be increased. The increase in energy is just sufficient to allow the average rate of energy input from field to free carriers to be passed on to the lattice via the emission of phonons. Loosely speaking, the temperature of the carriers is raised above the lattice temperature, thereby permitting a flow of energy from the carriers to the lattice. We say loosely speaking only because the energy distribution of the hot carriers is not, in general, a thermal distribution. We are concerned in this section with the mean energy of the hot carriers and discuss their distribution in energy separately below in the section on "Energy Distributions of Hot Carriers".

We make several approximations in order to simplify the calculations and to retain the emphasis on the physical processes taking place. These approximations still allow us to reproduce the first-order results already obtained in the literature by more detailed calculations.

The first approximation is that we neglect the energy spread of the carriers and treat them as if they all had the same energy. This is obviously a reasonable approximation for energy distributions that are sharply peaked at their mean energy.

The second approximation is that we confine the calculation to the dominant phonons in the spectrum of electron–phonon interaction. This means the shortest-wavelength phonons with which the electron can interact, namely, those with wavelengths comparable with the uncertainty radius of the electron. Since the interactions are strongly peaked at the shortest-wavelength phonons, the error involved is probably less than a factor of two.

Finally, we confine our treatment to "hot" as opposed to "warm" electrons. The mean energy of hot electrons is larger than about twice their thermal energy. The mean energy of warm electrons is only somewhat greater than their thermal energy. The distinction between the two regimes is roughly given by the following consideration. In the absence of an applied field, there is a certain rate of exchange of energy between electrons and phonons, the net flow in thermal equilibrium being, of course, zero. At low applied fields, the mean energy of the electron is increased slightly so that the balance is tipped slightly in favor of more energy flowing from electrons to phonons than the reverse. If, for example, the rate of emission of energy by an electron were proportional to its energy, one would expect only a 1% increase in mean energy of electrons when I^2R losses incurred by the applied field were 1% of the thermal equilibrium rate of exchange of energy between electrons and phonons. By the same argument, hot electrons would be generated when the energy delivered by the applied field becomes comparable with the thermal equilibrium rate of exchange. The mean energy of electrons would then be about twice their thermal energy. Other energy dependencies of the rate of emission of energy by electrons would yield qualitatively similar results.

For hot electrons, then, we can equate the rate of energy input by the field to the spontaneous rate of energy loss by the electrons (see Fig. 1)

$$\mathcal{E} e v_d = \frac{\hbar\omega}{\tau_e} = Av^n. \tag{27}$$

Here, v_d is the drift velocity ($\mathcal{E}\mu$) of electrons, τ_e the net emission time for phonons of frequency ω, and Av^n is the rate of emitting energy by electrons of energy $mv^2/2$. The complete expressions for the various phonons and couplings are given in Table I. By way of reminder, n is unity for nonpolar optical phonons and for acoustic phonons with piezoelectric coupling. n is 3 for acoustic phonons with deformation-potential coupling. For polar optical phonons, n is minus unity (ignoring the logarithmic factor) so that, in the approximation of a single energy for all electrons, there is no stable value for the energy of hot electrons above the energy of optical phonons. This instability is the basis for intrinsic dielectric breakdown (by impact ionization) as well as the basis for the Gunn effect.

We make use of the relations

$$v_d = \mathcal{E}\mu = \mathcal{E}\frac{\tau_c e}{m}, \tag{28}$$

$$2mv = \begin{cases} \dfrac{\hbar\omega}{v_s} \text{ for acoustic phonons } (v_s = \text{constant}) & \text{(29a)} \\ \dfrac{\hbar}{\lambda} \text{ for optical phonons } (\omega = \text{constant}) & \text{(29b)} \end{cases}$$

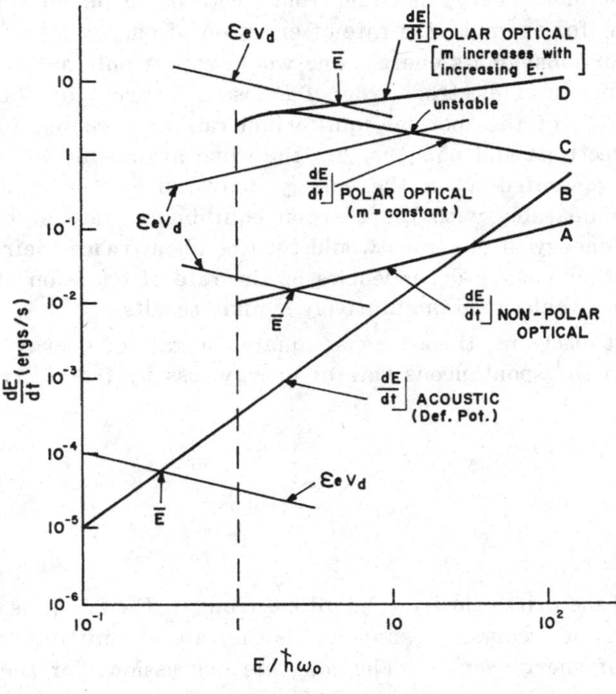

Fig. 1—Curves A, B, C, and D are representative rates of energy loss by electrons of energy E to the phonons indicated. Curves, labeled $\mathcal{E}ev_d$ are representative rates of energy gain by electrons of energy E from the applied field $\mathcal{E}$. The intersections marked $\bar{E}$ define the average energy of hot electrons.

$$\tau_e = \begin{cases} 2\dfrac{kT}{\hbar\omega}\tau_c \text{ for } 2kT > \hbar\omega & \text{(30a)} \\ \tau_c \quad \text{ for } 2kT < \hbar\omega & \text{(30b)} \end{cases}$$

where kT $(= mv_t^2/2)$ is the energy of electrons in thermal equilibrium with the lattice. The condition $2kT > \hbar\omega$ ensures that the induced emission and absorption due to thermal phonons is dominant over spontaneous emission. If spontaneous emission is dominant, $\tau_e = \tau_c$.

Interaction with Acoustic Phonons ($2kT > \hbar\omega$)

Combination of the left-hand side of Equation (27) with Equations (28), (29a), and (30a) yields

$$v = \frac{v_d v_t}{2v_s}. \tag{31}$$

Equation (31) states that the random velocity v of hot electrons is proportional to their drift velocity. It also defines the threshold field for the onset of hot electrons, namely, that field for which the drift velocity v_d is somewhat in excess of twice the velocity of sound.

Combination of Equations (27) and (31) yields

$$v = \left(\frac{2e}{A}\frac{v_s}{v_t}\right)^{1/(n-1)} \mathcal{E}^{1/(n-1)} \tag{32}$$

From Equations (31) and (32) the following derived relations are also of interest.

$$v_d = 2\left(\frac{2e}{A}\right)^{1/(n-1)} \left(\frac{v_s}{v_t}\right)^{n/(n-1)} \mathcal{E}^{1/(n-1)} \tag{33}$$

$$\mu \equiv \frac{v_d}{\mathcal{E}} = 2\left(\frac{2e}{A}\right)^{1/(n-1)} \left(\frac{v_s}{v_t}\right)^{n/(n-1)} \mathcal{E}^{(2-n)1/(n-1)} \tag{34}$$

For acoustic phonons coupled via a deformation potential, $n = 3$ (see Table I) and Equations (32)-(34) yield, in summary form,

$$\begin{aligned} &v \propto \mathcal{E}^{1/2} \\ &\mu \propto \mathcal{E}^{-1/2} \\ &v_d \propto \mathcal{E}^{1/2} \\ &l \equiv \text{mean free path} = \text{constant}. \end{aligned} \tag{35}$$

Since the energy ($mv^2/2$) of the hot electrons is proportional to $\mathcal{E}$ and since they begin to be heated when the drift velocity is about twice the velocity of sound, we can write

$$\text{mean energy of hot electrons} \equiv \bar{E} \approx \frac{\mathcal{E}}{\mathcal{E}_c} kT, \tag{36}$$

where $\mathcal{E}_c\mu \equiv 2v_S$. Equations (35) and (36) are in close agreement with Reference (6).

For acoustic phonons coupled via a piezoelectric field, $n = 1$. But for $n = 1$ all of the Equations (32)-(34) diverge. The meaning of this is that once the electrons begin to be heated, namely for fields greater than $2v_s/\mu$, there is no stable value for their energy at which the energy input from the field is balanced by energy loss to phonons. We can see this by noting, with the aid of Equation (31), that

$$\begin{aligned} \text{rate of energy input by field} &= \mathcal{E}\, e\, v_d \\ &= \mathcal{E}\, e \frac{2v_s}{v_t} v, \end{aligned}$$

$$\text{rate of energy loss to phonons} = Av.$$

Hence, when $2\mathcal{E}\, e\, (v_s/v_t) > A$, the electrons continue to receive more energy from the field than they can pass on to the phonons, and their mean energy continues to increase until some other type of phonon intervenes or until the assumptions leading to Equations (32)-(34) are no longer valid. The inequality $2\mathcal{E}\, e\, (v_s/v_t) > A$ occurs at the onset of hot electrons. This instability agrees with that analyzed by Laikhtman.[7]

Actually, energy loss to optical phonons will intervene to keep the energy of the hot electrons from diverging to large values. If the optical phonons did not intervene, the energy of the hot electrons would increase until Equation (30b) rather than Equation (30a) became valid. The condition for Equation (30b) is

$$\hbar\omega > 2kT,$$

or, since

$$\hbar\omega = 2mvv_s,$$

the energy of the hot electrons at the onset of Equation (30b) will then be

---

[6] W. Shockley, "Hot Electrons in Germanium and Ohm's Law," *Bell Syst. Tech. Jour.*, Vol. 30, p. 990, Oct. 1951.

[7] B. D. Laikhtman, "Deviations from Ohm's Law in Piezoelectric Semiconductors," *Soviet Physics-Solid State* (Translation), Vol. 6, p. 2573, May 1965.

$$\frac{mv^2}{2} > \left(\frac{kT}{mv_s^2/2}\right)\frac{kT}{4}. \tag{37}$$

At room temperature this energy is about 6 eV, and at liquid nitrogen about 0.3 eV.

Interaction with Acoustic Phonons ($\hbar\omega > 2kT$)

For $\hbar\omega > 2kT$, $\tau_e = \tau_c$ and Equation (27) becomes

$$\frac{\mathcal{E}^2 e^2 \tau_c}{m} = \frac{\hbar\omega}{\tau_c} = Av^n. \tag{38}$$

From the right-hand side of Equation (38), together with Equation (29a), we get

$$\tau_c = \frac{\hbar\omega}{Av^n} = \frac{2m\,v_s}{A}\,v^{1-n}. \tag{39}$$

From the left-hand side of Equation (38) we get

$$v_d \equiv \frac{\mathcal{E} e \tau_c}{m} = \left(\frac{\hbar\omega}{m}\right)^{1/2} = (2v\,v_s)^{1/2}. \tag{40}$$

Combination of Equations (39) and (40) yields the relations

$$v = \left(\frac{2e^2 v_s}{A^2}\right)^{1/(2n-1)} \mathcal{E}^{2/(2n-1)} \tag{41}$$

$$\mu \equiv \frac{(2v\,v_s)^{1/2}}{\mathcal{E}} = \left(\frac{e}{A}\right)^{1/(2n-1)} (2v_s)^{n/(2n-1)}\,\mathcal{E}^{(2-2n)/(2n-1)} \tag{42}$$

Equations (40)-(42) applied to acoustic phonons with deformation potential coupling, for which $n = 3$, give in summary form

$$\begin{aligned} v &\propto \mathcal{E}^{2/5} \\ \mu &\propto \mathcal{E}^{-4/5} \\ v_d &\propto \mathcal{E}^{1/5} \\ l &\propto v\mu \propto \mathcal{E}^{-2/5} \\ \bar{E} &\propto v^2 \propto \mathcal{E}^{4/5} \end{aligned} \tag{43}$$

The same equations applied to acoustic phonons with piezoelectric coupling ($n = 1$) give

$$
\begin{aligned}
v &\propto \mathcal{E}^2 \\
\mu &\propto \text{constant} \\
v_d &\propto \mathcal{E} \\
l &\propto v\mu \propto \mathcal{E}^2 \\
\bar{E} &\propto v^2 \propto \mathcal{E}^4
\end{aligned}
\tag{44}
$$

Note that the field dependence of the energy of hot carriers applies to the range of energies for which Equation (30b) is valid, namely,

$$E > \left(\frac{kT}{mv_s^2/2} \right) \frac{kT}{4}.$$

Interaction With Optical Phonons

The rate of energy loss to optical phonons rises abruptly as the mean carrier energy approaches the energy of an optical phonon.

The maximum rate of loss of energy to polar optical phonons occurs at a carrier energy of about $1.5\hbar\omega$ and then declines slowly, so that at a carrier energy of 10 eV the rate of loss is about ⅓ that of the maximum (see Figure 1).

The rate of loss to nonpolar optical phonons continues to increase after the abrupt rise and at a rate proportional to the velocity of the hot carrier.

When the mobility is dominated by optical phonons, Ohm's law is satisfied at low fields, and the drift velocity increases linearly with field until it approaches the value $(\hbar\omega/m)^{1/2}$. At this point, the applied field is sufficient to accelerate the carrier to the energy of an optical phonon in a time equal to its low-field-mobility mean free time. As the field is increased beyond this point, the drift velocity remains constant at the value $(\hbar\omega/m)^{1/2}$, the energy of the carrier remains constant at a value $\approx \hbar\omega$, and the mean free time varies inversely with applied field. These statements are derivable from Equation (38), for which $\hbar\omega > kT$ and $\tau_e = \tau_c$:

$$v_d \equiv \frac{\mathcal{E} e \tau_c}{m} = \left(\frac{\hbar\omega}{m} \right)^{1/2},$$

$$\mu \equiv \frac{v_d}{\mathcal{E}} = \left(\frac{\hbar\omega}{m} \right)^{1/2} \mathcal{E}^{-1},$$

$$v \approx v_d,$$
$$l \propto \mathcal{E}^{-1},$$
$$\bar{E} \approx \hbar\omega. \tag{45}$$

This pattern of behavior is known as "streaming", since the carrier is accelerated to an optical phonon energy in substantially a straight line, emits an optical phonon, and repeats the cycle. The pattern is maintained with increasing field until the time within which a carrier is accelerated to the energy of an optical phonon becomes shorter than the time for emitting an optical phonon. At this point, the mean energy of a carrier is increased beyond that of an optical phonon.

In the case of polar optical phonons, the energy of the carrier rapidly diverges with increasing field to energies that are limited by the advent of other sources of energy loss, such as impact ionization across the forbidden gap. This is the well known Fröhlich[8] model for intrinsic dielectric breakdown, as well as the model for the Gunn[9] instability. The critical field at which this instability sets in is obtained immediately by equating energy input to the maximum loss rate to optical phonons (see Table I):

$$\mathcal{E}_c \, e \, v_d = \beta \frac{e^2 \omega^2}{\epsilon_\infty v}. \tag{46}$$

Here, $$\beta = \frac{\epsilon_o - \epsilon_\infty}{\epsilon_0} \quad \text{and} \quad v \approx v_d = \left(\frac{\hbar\omega}{m}\right)^{1/2},$$

so that Equation (46) becomes

$$\mathcal{E}_c = \frac{\epsilon_o - \epsilon_\infty}{\epsilon_o \, \epsilon_\infty} \frac{e\omega m}{\hbar}. \tag{47}$$

Some representative values for $\mathcal{E}_c$ are given in Table II together with approximate values for critical fields taken from impact ioniza-

[8] H. Fröhlich, "Theory of Electrical Breakdown in Ionic Crystals," *Proc. Roy. Soc. London*, Vol. A160, p. 230, 18 May, 1937. For more recent discussion see R. Stratton, *Progress in Dielectrics*, Heywood and Co., Ltd., London, Vol. 3, p. 235, 1961.

[9] J. B. Gunn, "Instabilities of Current in III-V Semiconductors," *IBM Jour. Res. & Develop.*, Vol. 8, p. 141, April 1964.

Table II

Material	E_g (ev)	ϵ_0	ϵ_∞	$\frac{\epsilon_0 - \epsilon_\infty}{\epsilon_0 \epsilon_\infty}$	$\frac{m}{m_o}$	$\hbar\omega$ (ev)	$\mathcal{E}_c$ (V/cm) Computed (Equation 47)	Observed	Remarks
SiO_2	8	3.9	2.2	0.2	($\approx$1)	$\approx$0.1	4.5×10^6	5×10^6	dielectric breakdown
NaCl	8	5.7	2.2	0.3	($\approx$1)	0.024	2×10^6	1.6×10^6	dielectric breakdown
CdTe	1.4	10.6	7.2	4×10^{-2}	0.1	0.02	2×10^4	10^4	Gunn effect
GaAs	1.4	12.5	11	10^{-2}	0.07	0.033	5×10^3	3×10^3	Gunn effect
InAs	0.36	14	11.6	1.5×10^{-2}	0.02	0.03	2×10^3	10^3	impact ionization
InSb	0.17	17	16	3.7×10^{-3}	0.015	0.025	3×10^2	4×10^2	impact ionization
CdS	2.5	9.3	5.2	8×10^{-2}	0.2	0.036	1.3×10^5	2×10^6	dielectric breakdown
ZnSe	2.6	8.1	5.9	4.5×10^{-2}	0.2	0.03	6×10^4	2×10^6	impact ionization
ZnO	3.3	8.5	3.7	0.15	0.3	0.07	7×10^4	4×10^6	impact ionization
GaP	2.2	10.2	8.5	2×10^{-2}	0.12	0.05	2.8×10^4	$\approx 10^6$	impact ionization
GaAs	1.4	12.5	11	10^{-2}	0.5	0.033	3.5×10^4	5×10^5	impact ionization
GaAs-GaP	2.0							10^6	impact ionization

Comments:

(1) It is not certain that dielectric breakdown in SiO_2 and NaCl is caused by impact ionization. The computed values only show that impact ionization is not expected to occur before the observed dielectric breakdown. The value $m/m_o = 1$ is chosen as a nominal value for computational purpose.

(2) For CdTe, GaAs, InAs and InSb, the concurrence of computed and experimental values for $\mathcal{E}_c$ is significant. Both the Gunn effect and impact ionization in small-bandgap materials emphasize the band properties near the band edge.

(3) For the remaining materials, the large discrepancy between observed and computed values of $\mathcal{E}_c$ is strong evidence that impact ionization is determined by the band properties at energies several volts above the band edge. Increased effective mass is one major contribution. Additional energy loss by intervalley scattering (e.g., in GaAs, GaP and GaAs-GaP) is a second.

References for Table II

SiO_2—A. M. Goodman, private communication; N. Klein and H. Gafni, "The Maximum Dielectric Strength of Thin Silicon Oxide Films," *IEEE Trans. on Electron Devices,* Vol. ED-13, p. 281, Feb. 1966.

NaCl—R. Cooper, "The Electric Breakdown of Alkali Halide Crystals," *Progress in Dielectrics,* Academic Press, Inc., New York, Vol. 5, p. 97, 1963; and R. Williams, "High Electric Fields in Sodium Chloride," *Jour. Phys. Chem. Solids,* Vol. 25, p. 853, 1964.

CdTe—M. R. Oliver, A. L. McWhorter, and A. G. Foyt, "Current Runaway and Avalanche Effects in n-CdTe," *Appl. Phys. Letters,* Vol. 11, p. 111, 15 Aug. 1967.

GaAs—Gunn Effect—State of the Art.

InAs—M. Steele and S. Tosima, "Electron-Hole Scattering in Solids Exhibiting Band-Gap Impact Ionization," *Japanese Jour. Appl. Phys.,* Vol. 2, p. 381, July 1963; J. W. Allen, M. Shyam, and G. L. Pearson, "Gunn Oscillations in Indium Arsenide," *Appl. Phys. Letters,* Vol. 9, p. 39, 1 July 1966.

InSb—M. Toda, "A Plasma Instability Induced by Electron-Hole Generation in Impact Ionization," *Jour. Appl. Phys.,* Vol. 37, p. 32, Jan. 1966; J. C. McGroddy and M. I. Nathan, *Jour. Phys. Soc. Japan,* Vol. 21, Suppl. 437, 1966.

CdS—R. Williams, "Dielectric Breakdown in Cadium Sulfide," *Phys. Rev.,* Vol. 125, p. 850, Feb. 1, 1962; A. Many, "High-Field Effects in Photoconducting Cadium Sulphide," *Jour. Phys. Chem. Solids,* Vol. 26, p. 575, 1965; K. W. Böer and K. Bogus, *Phys. Rev.,* Vol. 176, p. 899, 1968.

ZnSe—R. Williams, "Impact Ionization in ZnSe and Comparison with CdS," *Physics Letters,* Vol. 25A, p. 445, 25 Sept. 1967.

ZnO—R. Williams and A. Willis, "Electron Multiplication and Surface Charge on Zinc Oxide Single Crystals," *Jour. Appl. Phys.,* Vol. 39, p. 3731, July 1968.

GaP—H. G. White and R. A. Logan, "GaP Surface-Barrier Diodes," *Jour. Appl. Phys.,* Vol. 34, p. 1990, July 1963.

GaAs—H. Kressel and G. Kupsky, "The Effective Ionization Rate for Hot Carriers in GaAs," *Int. Jour. Electronics,* Vol. 20, p. 535, June 1966.

GaAs-GaP—R. Williams, "Impact Ionization and Charge Transport in GaAs:GaP 50% Alloy," *Jour. Appl. Phys.,* Vol. 39, p. 57, Jan. 1968.

tion, dielectric breakdown, or the Gunn instability. Factor of two agreement is confirmed in the upper half of Table II from critical fields of 10^2 to 10^6 volts/cm. A word of caution is, however, in order for NaCl, or the alkali-halides in general. While extensive work has been carried out in the last forty years on dielectric breakdown in the alkali-halides, and while the results appeared to bear out Fröhlich's model, it is highly likely that breakdown in these materials is caused by field emission from the electrodes together with heating due to large space-charge-limited currents rather than by impact ionization.[10] If the field emission could, however, be avoided, it is likely that the breakdown by the Fröhlich mechanism would take place at very nearly the same fields. In the smaller gap materials, the critical field is smaller and there is less likelihood of competitive processes such as field emission.

In the case of nonpolar optical phonons, we return to Equation (38) using the value $n = 1$ obtained from Table I. From the right-hand side of Equation (38) we get

$$v = \frac{\hbar\omega}{\tau_c A}, \tag{48}$$

and from the left-hand side,

$$\tau_c = \frac{1}{\mathcal{E}e} (\hbar\omega m)^{1/2}. \tag{49}$$

Equations (48) and (49) combine to yield

$$v = \frac{e}{A}\left(\frac{\hbar\omega}{m}\right)^{1/2} \mathcal{E},$$

$$\mu = \left(\frac{\hbar\omega}{m}\right)^{1/2} \mathcal{E}^{-1},$$

$$v_d = \left[\frac{\hbar\omega}{m}\right]^{1/2}$$

$$l = \text{constant}. \tag{50}$$

---

[10] R. Williams, "High Electric Fields in Sodium Chloride," *Jour. Phys. Chem. Solids*, Vol. 25, p. 853, 1964.

Here, the carrier energy increases as $\mathcal{E}^2$, while the drift velocity remains constant and the mobility and mean free time decrease as $\mathcal{E}^{-1}$. The energy of hot carriers can be written nominally in the form

$$\bar{E} \approx \left(\frac{\mathcal{E}}{\mathcal{E}_c}\right)^2 \hbar\omega, \tag{51}$$

where $\mathcal{E}_c$ is the field at which an electron is accelerated to the energy of an optical phonon in a time equal to the time required for an electron to emit an optical phonon. Thus,

$$\mathcal{E}_c \frac{e}{m} \tau_e = \left(\frac{2\hbar\omega}{m}\right)^{1/2}$$

or

$$\mathcal{E}_c = \frac{(2\hbar\omega m)^{1/2}}{e\tau_e} = \frac{2\hbar\omega}{el}. \tag{52}$$

If we take l to be about 10^{-6} cm (see discussion in the section on Rates of Impact Ionization) and $\hbar\omega = 0.06$ eV (values appropriate to silicon),

$$\mathcal{E}_c = 10^5 \text{ V/cm}$$

Significant impact ionization should then take place at fields sufficient to yield a mean energy of hot electrons of the order of half the energy of the forbidden gap. By Equation (51), this would mean

$$\mathcal{E} \approx 3 \times 10^5 \text{ volts/cm},$$

a value that compares favorably with experimental values around 2×10^5 volts/cm.[11]

ENERGY DISTRIBUTIONS OF HOT CARRIERS

General Remarks

In the previous section the mean energies of hot carriers were obtained by simplifying assumptions that kept the physical processes in constant view. The results match those obtained by a variety of

---

[11] C. A. Lee, R. A. Logan, R. L. Batdorf, J. J. Kleimack, and W. Wiegmann, "Ionization Rates of Holes and Electrons in Silicon," *Phys. Rev.*, Vol. 134, p. A761, 4 May 1964.

authors by more detailed considerations—usually carried out in momentum space rather than in real space. We extend the analysis to obtain some insight into the distribution in energy about the mean energy,* again by using assumptions designed to keep the physical processes in evidence and the analysis compact. It must be remembered that there are still remarkably little, if any, direct data on any actual distributions, and that these are difficult to obtain.** The indirect estimates of energy distributions based, for example, on interpreting rates of impact ionization in back-biased junctions are tenuous indeed.

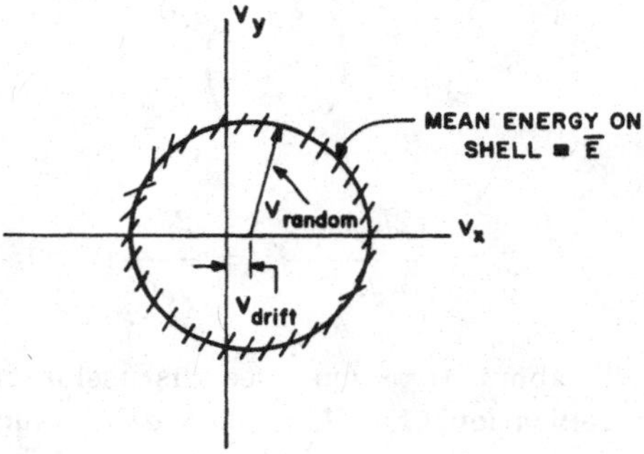

Fig. 2—A drifted Maxwellian distribution showing schematically the large ratio of random velocity v_r to drift velocity v_d.

Basis for Analytical Model

The problem to be analysed is sketched in Figure 2. We have a distribution of electrons in momentum space located predominantly on a roughly spherical shell that is displaced from the origin by the drift velocity. The random velocities are, in general, large compared with the drift velocity. A major exception is the case of streaming, discussed in the previous section under optical phonons, for relatively low-energy ($\approx \hbar\omega_{\text{optical}}$) electrons. A less pronounced departure is an elongation of the spherical distribution in the direction of drift. The major effect of these departures is to increase the mean drift velocity and hence the mean power input $\mathcal{E}ev_d$.

We wish to estimate the distribution of electrons in energy shells

* The emphasis in the present paper is on electron energies in excess of the optical phonon energy. For a treatment of energies below the optical phonon energy, see, e.g., R. S. Crandall, *Phys. Rev.*, Vol. 169, p. 585 (1968).

** See, e.g., E. D. Savoye and D. E. Anderson, *Jour. Appl. Phys.*, Vol. 38, p. 3245 (1969).

neighboring the shell of mean energy $\bar{E}$. For most problems of an applied nature, the distributions of electrons for energies greater than $\bar{E}$ is of major interest. The physical processes determining $\bar{E}$ are sketched in Figure 3. Here we show the dominant energy input, $\mathcal{E}ev_d = \mathcal{E}^2e^2\tau_c/m$, which forces electrons toward higher energy shells. A second curve gives the dominant energy loss $\hbar\omega/\tau_e$, which forces electrons towards lower energies. These two curves balance each other at the mean energy $\bar{E}$. The input curve is shown decreasing toward higher energies, while the loss curve is shown increasing towards higher energies. These slopes are based on the fact that in general,

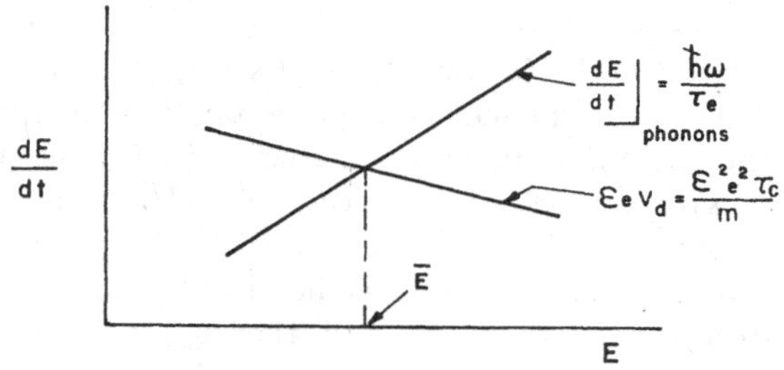

Fig. 3—Representative rates of energy gain and loss curves as a function of electron energy E. The intersection at $\bar{E}$ defines the mean energy of hot electrons.

τ_c and τ_e decrease toward higher energies. If all of the electrons were initially concentrated at $\bar{E}$, they would tend to spread out towards higher and lower energy shells by a kind of diffusion process resulting from the statistics of random emission and absorption of phonons. Our analytic model does in fact, calculate the energy distribution on the high side of $\bar{E}$ by equating the diffusive spreading force to the average loss rate $\hbar\omega/\tau_e$.[12]

Figure 3 is a plot in one-dimensional energy space, whereas the primary distribution in Figure 2 is in three-dimensional momentum space. The transformation would certainly be valid if the rate of jumping between opposite surfaces of the sphere were fast compared with the average drift under the action of either of the force curves. In this way an electron would tend to sample various parts of the sphere on the average as its drifts to or from the shell $\bar{E}$. Interaction with

[12] See also L. V. Keldysh, "Concerning the Theory of Impact Ionization in Semiconductors," *Soviet Phys. JETP* (Translation), Vol. 21, p. 1135, Dec. 1965; and T. Kurosawa, "Notes on the Theory of Hot Electrons in Semiconductors," *Jour. Phys. Soc. Japan*, Vol. 20, p. 937, June 1965 for more detailed treatments of diffusion in energy space.

acoustic phonons satisfies this condition easily, since the collisions are predominantly elastic (on the average, $2n_{Ph}+1$ collisions are incurred per net loss of one phonon). n_{Ph} is the phonon occupancy, $kT/\hbar\omega \approx v_t/v_s$. In the case of nonpolar optical phonons, the condition is only marginally satisfied since $\tau_e = \tau_c$, that is, the rate of sampling different parts of the sphere is the same as the net rate of emission of phonons. This would tend to populate the sphere more heavily on the right-hand surface, that is, in the direction of the average drift. Such a distortion would tend to increase the average drift velocity and lead to poor saturation of the drift velocity in the hot-electron range. The fact that the drift velocity for n-type Ge saturates very well and for Si reasonably well (within a factor of two) is evidence that the distortion does not play a major role.

There is, however, one factor that we have ignored in transforming from three-dimensional momentum space to one-dimensional energy space. A particle in three-dimensional space will diffuse away from its point of origin with a velocity of the order of $v_r(l/r)$ where v_r is its random velocity, l its mean free path, and r its distance from the origin. This diffusive force stems from the geometrical fact that spherical shells at larger radii have larger volumes and, thereby, increase the probability that a randomly diffusing particle will gravitate toward larger radii. In the present case v_r, r and l refer to velocity, radius, and mean free path in velocity space. This means that, in the absence of spontaneous emission, an electron will drift toward larger energies. This drift should properly be added in Figure 3 to the term $\mathcal{E}ev_d$, forcing an electron to the right. However, as shown in the appendix, the correction is comparable with $\mathcal{E}ev_d$, which we already neglect on the high side of $\bar{E}$ since (as seen by inspection of Figure 3) it rapidly becomes small compared with $\hbar\omega/\tau_e$.

For the purpose of calculating the high-energy tail of electron energies, we have stripped our problem down to the simple balance between diffusive flow in energy space and the energy loss rate due to spontaneous phonon emission;

$$-D\frac{dn}{dE} = n\frac{\hbar\omega}{\tau_e}, \tag{53}$$

where n is the density of electrons in energy space. The diffusion constant D is, generally,

$$D = \langle v^2\tau_c \rangle. \tag{54}$$

where v is a random velocity and τ_c the mean free time between collisions. The random velocity in energy space is the energy change suffered by the electron moving stochastically with or against the field divided by the time between phonon collisions. Thus,

$$D = \left(\frac{\mathcal{E}el}{2\tau_c} \right)^2 \tau_c, \tag{55}$$

where l is the mean free path between collisions. The factor 2 comes from averaging the projection of l along the field. Equation (53) then becomes

$$-\frac{(\mathcal{E}el)^2}{4\tau_c}\frac{dn}{dE} = n\frac{\hbar\omega}{\tau_e}. \tag{56}$$

High-Energy Distribution for Interaction with Acoustic Phonons

We use Equation (56) together with the relations

$$\frac{1}{2}mv^2 = E,$$

$$\tau_e = \frac{2kT}{\hbar\omega}\tau_c,$$

$$\hbar\omega = 2mvv_s,$$

to obtain

$$n = n_o \exp\left[\frac{-8mv_s{}^2E^2}{(\mathcal{E}el)^2kT}\right], \tag{57}$$

and the further relations*

$$l = v_t\tau_c,$$

$$\mu = \tau_c\frac{e}{m},$$

* Note: l is a constant for acoustic phonons; v_t is the thermal equilibrium velocity, and τ_c is the low-field collision time.

$$\mathcal{E}_o = \frac{v_s}{\mu},$$

to obtain

$$n = n_o \exp\left[-\left[\frac{E}{\dfrac{\mathcal{E}}{2\mathcal{E}_o}kT}\right]^2\right]. \tag{58}$$

But the mean energy of hot electrons for acoustic phonons coupled via a deformation potential is (see Equation (36))

$$\bar{E} = \frac{\mathcal{E}}{2\mathcal{E}_o}kT.$$

Hence, the final result is

$$n = n_o \exp\left[-\left(\frac{E}{\bar{E}}\right)^2\right], \tag{59}$$

namely, a distribution more sharply peaked near $\bar{E}$ than would be given by the normal temperature distribution exp $(-E/\bar{E})$.

The distribution obtained by Yamashita and Watanaba[13] by a more careful treatment has the same form as Equation (57).

Energy Distributions for Interaction with Nonpolar Optical Phonons

Again, we begin with the Equation (56) and, recognizing that $\tau_e = \tau_c$, we obtain at once

$$n = n_o \exp\left[-\frac{4\hbar\omega E}{(\mathcal{E}el)^2}\right], \tag{60}$$

where, by way of reminder, the mean free path l is a constant. From Equations (51) and (52),

---

[13] J. Yamashita and M. Wantanabe, "On the Conductivity of Non-Polar Crystals in the Strong Electric Field, I," *Prog. Theor. Phys.*, Vol. 12, p. 443, Oct. 1954.

$$\bar{E} = \frac{(\mathcal{E}el)^2}{4\hbar\omega}, \tag{61}$$

and Equation (60) can then be written

$$n = n_o \exp\left(-\frac{E}{\bar{E}}\right). \tag{62}$$

In brief, the energy distribution for energies greater than $\bar{E}$ does, in this case, follow a normal temperature distribution. Again a more detailed analysis by Wolff[14] yields Equation (60), with a factor of 3 in place of 4 in the argument.

At fields for which $\mathcal{E}el < \hbar\omega$, $\bar{E} < \hbar\omega$ and Equation (61) loses its significance. At these low fields, the energy distribution of electrons is controlled mainly by statistical "leakage" through the optical phonon barrier. The probability that an electron will avoid emitting an optical phonon for n emission times is $\exp(-n)$. The energy acquired by such an electron will be

$$E = n\,(\mathcal{E}el). \tag{63}$$

Hence the energy distribution of electrons at these low fields is

$$n \approx n_o \exp\left(-\frac{E}{\mathcal{E}el}\right). \tag{64}$$

Multiple scattering processes have been ignored in writing Equation (63). These will tend to increase the "leakage" past the optical phonon barrier.

Energy Distribution for Interaction with Polar Optical Phonons

It was meaningful to compute a mean energy and an energy distribution for hot electrons interacting with acoustic phonons and with nonpolar optical phonons, because the rates of energy loss to these phonons increase with increasing electron energy. In brief, the distribution is stable (see Figure 1). In contrast, the rate of energy loss to polar optical phonons decreases with increasing electron energy for energies above that of one or two optical phonons. One can expect

[14] P. A. Wolff, "Theory of Electron Multiplication in Silicon and Germanium," *Phys. Rev.*, Vol. 95, p. 1415, Sept. 15, 1954.

certain instabilities to set in as the applied field is increased. And, indeed, if all of the electrons had the same energy, the instability would set in quite abruptly (see Figure 1) when the applied field delivered more energy to the electrons than the maximum rate at which they could lose energy to the phonons. This has been the model for intrinsic dielectric breakdown of ionic insulators—a phenomenon that is known to occur abruptly as the electric field is increased. While the mechanism for breakdown in high-bandgap insulators, particularly in alkali halides for which the model was designed, is still in some doubt, it is nevertheless true that the penetration of the optical phonon barrier does occur abruptly for materials such as GaAs, GaP, and CdTe, as observed in the Gunn effect.

For applied fields up to within a factor of two of the breakdown field, the mean energy of electrons should not significantly exceed that of an optical phonon, namely, about 0.05 eV. For fields at and above breakdown, the energy of the electrons should diverge toward infinity or until interrupted by a new loss mechanism. The latter is commonly impact ionization across the forbidden gap (as in the case of dielectric breakdown) or transfer to a higher mass band as in GaAs. In any event, significant hot-electron energies occur within a narrow range of electric fields near breakdown. There have been attempts[15] to describe analytically the mean energy or "temperature" of electrons in this range. Such analyses must be of questionable value, owing both to the approximations that are normally made to get a tractable analysis and to the difficulty of experimentally confirming any particular analytic results.

The dominant feature of the hot-electron energy is its rapid divergence from a few hundreths of an electron volt toward infinity within about a factor of two in electric field. What we present here is an approximate description of that divergence as derived from Equation (56). The fact that there is even a finite range of field that can be discussed analytically depends on the electrons having a finite energy spread as opposed to being located all at the same energy. The latter would lead to a step-function divergence as a function of field, as mentioned above. Hence, the following simplified analysis yields a form for the energy distribution.

We use an approximate form for the energy gain and loss curves as shown in Figure 4. The loss curve rises abruptly from zero to its maximum value at $E = \hbar\omega$. For higher energies it remains constant.

---

[15] R. Stratton, "The Influence of Interelectronic Collisions on Conduction and Breakdown in Polar Crystals," *Proc. Roy. Soc. London*, Vol. 246A, p. 406, 19 Aug. 1958.

The actual curve has a broad maximum and a slow decrease at higher energies. We take also the energy input curve to be substantially constant as a function of electron energy $\bar{E}$. Both these assumptions mean that $\tau_e = \tau_c =$ constant. Equation (56) then becomes

$$-\frac{(\mathcal{E}el)^2}{4\tau_c}\frac{dn}{dE} = n\left[\frac{\hbar\omega}{\tau_c} - \frac{\mathcal{E}^2e^2\tau_c}{m}\right]. \tag{65}$$

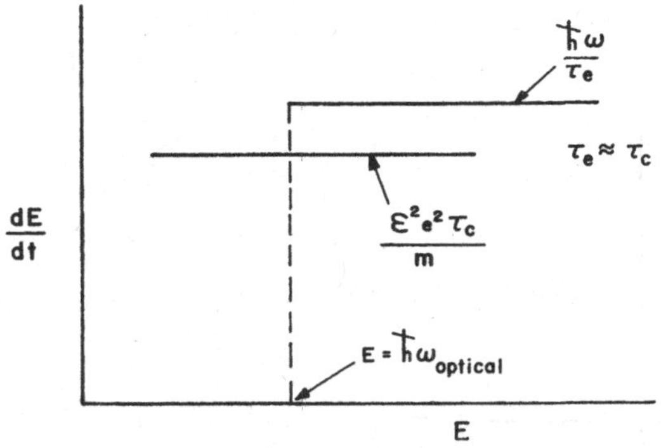

Fig. 4—Schematic approximation for rates of energy gain and loss for polar optical phonons for an electric field approaching breakdown.

For $l = v\tau_c$, $\tau_c =$ constant, and $mv^2/2 = E$, Equation (65) is readily integrable in the form

$$n = n_o\left(\frac{E}{E_o}\right)^{-2\,[(m\hbar\omega/(\mathcal{E}e\tau_c)^2) - 1]}$$

or

$$n = n_o\left(\frac{E}{E_o}\right)^{-2\,\{[(\hbar\omega/\tau_c)/\mathcal{E}ev_d] - 1\}} \tag{66}$$

n_o is the density of electrons at energy $E_o \approx \hbar\omega$. Equation (66) states immediately that when the rate of energy input $\mathcal{E}ev_d$ is equal to the rate of loss $\hbar\omega/\tau_c$, the electron energy diverges to infinity. This is the criterion for breakdown. If we take as an approximate measure for the average energy ($\bar{E}$) the value of E at which $n = n_o/2$, we can plot

$\bar{E}$ as a function of applied field $\mathcal{E}$. The result in Figure 5 shows that even at a field as high as half the breakdown field, the average energy of the electrons is only $\sqrt{2}\,\hbar\omega$. The calculation of Hilsum[16] based on Stratton's expression for electron temperature shows, if anything, an even more abrupt divergence at the breakdown field. That is, the field at which the electrons achieve an energy of $\sqrt{2}\hbar\omega$ lies even closer to the breakdown field than the value of ½ we calculate here.

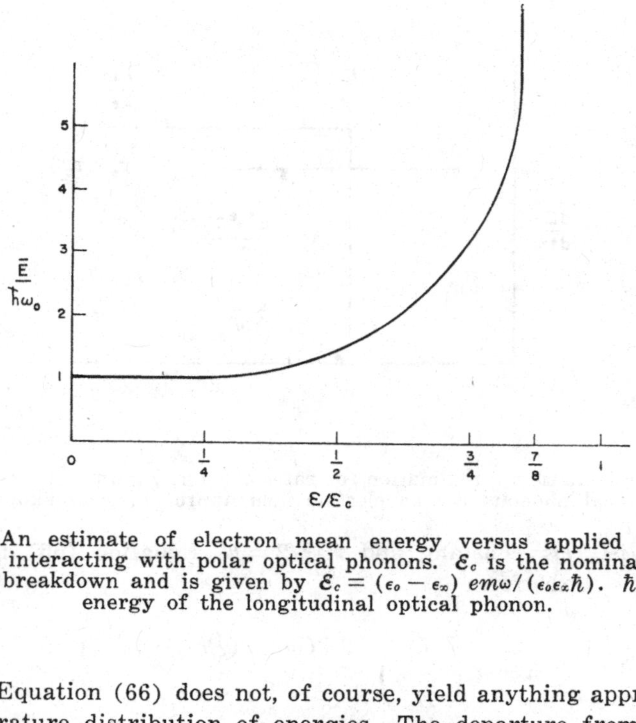

Fig. 5—An estimate of electron mean energy versus applied field for electrons interacting with polar optical phonons. $\mathcal{E}_c$ is the nominal electric field for breakdown and is given by $\mathcal{E}_c = (\epsilon_o - \epsilon_\infty)\, em\omega/(\epsilon_o\epsilon_\infty\hbar)$. $\hbar\omega_o$ is the energy of the longitudinal optical phonon.

Our Equation (66) does not, of course, yield anything approaching a temperature distribution of energies. The departure from such a distribution is even more marked when one recognizes that even the distribution of Equation (66) must be cut off at electron energies such that $\mathcal{E}ev_d > \hbar\omega/\tau_c$. This is shown in Figure 1 on curve C where the two curves have their second crossing. These high-energy electrons continually gain energy from the field until they encounter a new energy-loss process.

Equation (66) does show that for larger bandgap energies, the dependence of the rate of impact ionization on electric field becomes

[16] C. Hilsum, "Transferred Electron Amplifiers and Oscillators," *Proc. I.R.E.*, Vol. 50, p. 185, Feb. 1962.

progressively steeper. For impact ionization, the value of E in Equation (66) is chosen to be comparable with the forbidden gap energy.

Rate Of Impact Ionization

An analysis of hot electrons should define the field at which significant impact ionization takes place or, in more detail, the rate of impact ionization as a function of field. The field at which significant impact ionization occurs is relatively easy to estimate. It is that field at which the mean energy of the hot carriers is of the order of a half or a quarter of the energy needed for impact ionization. The spread in energy of the actual distribution around this mean energy will then provide a significant fraction of carriers with energies sufficient to cause impact ionization. Equations (36) and (61) relate the mean energy to the electric field and readily yield estimates of the required fields, provided, of course, that the mean free path for phonon emission is known. These equations apply to interaction with acoustic phonons and with nonpolar optical phonons, respectively. For polar optical phonons, the critical field is given by Equation (47).

Nonpolar Optical Phonons

The rate of impact ionization as a function of electric field should be readily derivable from any of the energy distributions given by Equations (58), (60), (64), or (66). That is, one inserts into these equations a value for E (the energy at which significant impact ionization takes place) and a value for l (the mean free path for phonon emission) and proceeds to calculate n/n_0, a measure of the fraction of electrons with energies greater than E, as a function of electric field. Alternatively, the same procedure can be used to compute a value for l knowing experimentally the rate of impact ionization as a function of field. This is, indeed, what has been done for impact ionization in back-biased p-n silicon junctions. References to this literature are to be found, for example, in the key paper by C. A. Lee et al[11] on ionization measurements in such junctions. Also cited in this paper is the long controversy as to whether the energy distribution used to compute the rate of ionization should be of the form exp $(-A/\mathcal{E}^2)$ shown in Equation (60) and associated with Wolff[14] or of the form exp $(-A/\mathcal{E})$ shown in Equation (64) and associated with Shockley.[17] Finally, the paper cites the resolution of the controversy by the more detailed calcu-

[17] W. Shockley, "Problems Related to p-n Junctions in Silicon," *Solid-State Electronics*, Vol. 2, No. 1 p. 35, 1961.

lations by Baraff,[18] resulting in a mixture of these two forms. In sum, there appears to be general agreement that experimental data should be interpreted in terms of Baraff's universal curves in order to define a proper value for the mean free path for phonon emission.

The major point we wish to make here is that the difference between the models of Baraff and his predecessors, Wolff and Shockley, is small compared with the difference between all three authors and the actual situation. The reason for this statement is that all of the authors have assumed a constant mean free path, l, independent of energy. This is a proper assumption if the effective mass also is a constant. In fact, however, the effective mass is expected to vary in silicon from about 0.3 near the band edge to a value approaching unity at energies of a volt or more above the band edge. The precise variation is not important here. Since the mean free path depends inversely on the *square* of the effective mass, the mean free path will decrease by almost a factor of ten in the energy range of interest. All of the calculations ignore this variation of mean free path by a *factor of ten* and, at the same time, produce results that are used to deduce a mean free path from experimental data supposedly accurate to about *ten per cent.* This mismatch between the models on which the several analyses have been based and the actual state of affairs, while bad enough, is not quite so serious as the factor of ten would suggest. All of the analyses tend to emphasize the mean free path at energies near the ionization energy rather than near the band edge. This comes about because the average energy $\bar{E}$ depends on the local, that is, high-energy mean free path, at least to the extent that the energy distribution is strongly peaked near $\bar{E}$.

Consistent with the approximate nature of the models for hot-electron energy distributions, we suggest here a simple and approximate method for interpreting experimental data. We note, for example that the data on impact ionization (see Lee et al[11]) cover only a factor of two in electric field. Over this factor of two, the rate of ionization by electrons increases by a factor of 100. The dominant fact is the steep rise of the rate of ionization with field, and not the particular shape of the rise curve. The latter is a tenuous hook on which to hang the validity of a theory. The question we ask, then, is what value of mean free path will yield a 100-fold increase in rate of ionization for a twofold increase in field at the field in question? In terms of Equation (60), and using the more accurate factor of 3 in the exponent,

---

[18] G. A. Baraff, "Distribution Functions and Ionization Rates for Hot Electrons in Semiconductors," *Phys. Rev.*, Vol. 128, p. 2507, Dec. 15, 1962. This dependence stems from the dependence of the coupling constant β on effective mass via the relation $\hbar/\lambda = 2\, m v v_s$.

$$\frac{\exp\left[-\dfrac{3\hbar\omega E_i}{(2\mathcal{E}el)^2}\right]}{\exp\left[-\dfrac{3\hbar\omega E_i}{(\mathcal{E}el)^2}\right]} = 10^2,$$

or

$$2.2\frac{\hbar\omega E_i}{(\mathcal{E}el)^2} = 4.6.$$

The relative importance of the several parameters is clearly assessable in this expression. The right-hand side is ln (100) and measures the steepness of the ionization curve. Using the values taken from Lee et al, $\hbar\omega = 0.06$ eV, $E_i = 1.6$ eV, and $\mathcal{E} = 2 \times 10^5$ V/cm, we compute $l = 110$ Å.

A similar computation using Equation (64) is expressed in the form

$$\frac{\exp\left(-\dfrac{E_i}{2\mathcal{E}el}\right)}{\exp\left(-\dfrac{E_i}{\mathcal{E}el}\right)} = 10^2,$$

and yields a value of 90 Å for l. These values are to be compared with 70 Å computed by Lee et al using Baraff's curves. It is worth noting that Equations (60) and (64) do not lead to widely different values of l.

Another simple estimate of l can be made by equating the average energy of electrons to, say, half the ionization energy at the highest fields for which measurements could be made before breakdown. This estimate favors the local high energy value of l.

$$\frac{(\mathcal{E}el)^2}{4\hbar\omega} = \frac{1.6}{2}.$$

For $\mathcal{E} = 4 \times 10^5$ V/cm, $l = 110$ Å.

One note to be kept in mind is that a mean free path of 100 Å yields an emission time of less than 10^{-14} sec, i.e., less than the ω^{-1} of the emitted phonons, and raises questions about the validity of perturbation theory.

Polar Optical Phonons

The field dependence of the rate of impact ionization in the case of electron interaction with *polar* optical phonons should be experimentally too difficult to define for materials with forbidden gaps of a volt or more, owing to the runaway character of the hot electrons. From Equation (66), the field required to produce an average energy of electrons of about a volt is within a few percent of the breakdown field. It would be difficult, indeed, to ensure such uniform ionization over a specimen cross section. These remarks are based on the assumption of a constant effective mass.

It is true that the field dependence of the rate of ionization has been successfully observed for InSb[19, 20] and analysed by Dumke[21]. The forbidden gap, however, is only 0.2 volt, and is sufficiently close to the energy of an optical phonon to permit statistical leakage of electrons past the optical phonon barrier at fields well removed from breakdown. Even here the rate of ionization increases at about the tenth power of the electric field. McGroddy and Nathan[20] computed a mean free path for phonon emission of 1 micron using Baraff's model for *nonpolar* optical phonons. Dumke computed substantially the same values using his model for *polar* optical phonons. Such universal agreement can only be read as the insensitivity of this parameter to "modeling." The dominant factor, as we have pointed out, for computing the mean free path for phonon emission is the electric field at which significant impact ionization takes place. The mean free path is also readily derivable from the material parameters, as shown in Chap. IV. There, for example, the mean free path for optical phonon emission was given as

$$l = r_H \frac{\epsilon_0 \epsilon_\infty}{\epsilon_0 - \epsilon_\infty} \frac{m_0}{m},$$

where r_H = radius of the Bohr orbit (0.5Å), $\epsilon_0\epsilon_\infty/(\epsilon_0 - \epsilon_\infty) \approx 2 \times 10^2$ and $m_0/m \approx 50$ for InSb. With these parameters l is again computed to be about 10^4 Å. For the alkali halides, at the other extreme, l is only a few angstroms.

---

[19] M. Toda, "A Plasma Instability Induced by Electron-Hole Generation in Impact Ionization," *Jour. Appl. Phys.*, Vol. 37, p. 32, Jan. 1966.

[20] J. C. McGroddy and M. I. Nathan, *Jour. Phys. Soc. Japan*, Vol. 21, Suppl. 437, 1966.

[21] W. P. Dumke, "Theory of Avalanche Breakdown in InSb and InAs," *Phys. Rev.*, Vol. 167, p. 783, 15 March, 1968.

While we have emphasized the highly precipitous and runaway character of the mean energy of hot electrons versus applied field in polar materials, it should be kept in mind that these arguments are based on a reasonably constant effective mass. If, on the other hand, the effective mass increases by a factor of three or more in the range of a few volts above the band edge, the loss curve dE/dt should increase (rather than decrease) with increasing energy (see Figure 1, Curve D). Under these conditions, an increasing electric field will lead to stable distributions of hot electrons whose mean energy versus field can readily be estimated in terms of the slope of the dE/dt curve. Alternatively, if at higher energies a new phonon loss mechanism is introduced, such as inter-valley scattering, the runaway character can be arrested and a stable high-energy-electron distribution maintained. Either or both these mechanisms can account, as Williams[22] has pointed out, for the relatively high fields required to impact ionize in CdS, ZnSe, ZnO, GaAs and GaP (see lower half of Table II).

Dielectric Breakdown

Dielectric breakdown by impact ionization will occur when an electron on the average creates one successor by impact ionization during its lifetime. The condition can be expressed approximately by

$$\frac{L_D}{L_I}\,\frac{n(E_I)}{n_o} = 1,$$

where L_D is the drift length of an electron during its lifetime, L_I the distance, in the field direction, an electron of energy E_I travels between impact ionizations, and $n(E_I)/n_o$ is the fraction of electrons having an energy E_I at which significant impact ionization takes place. E_I is somewhat larger than the energetic threshold for ionization. Since $n(E_I)/n_o$ is an exponential function of electric field, the breakdown field strength will depend only weakly (logarithmically) on the lifetime of an electron. Morever since $n(E_I)/n_o$ is such a steep function of electric field, one can conclude that the mean energy $\bar{E}$ of electrons will be in the order of half or more of E_I. This value of $\bar{E}$ then, defines an electric field (see Equations (36), (47), and (61) that should be within a factor of two of the breakdown field. A closer examination of breakdown field strength is hardly warranted, unless the fine details of band

[22] R. Williams, "High Electric Fields in II-VI Materials," *Applied Optics*, Supplement 3, p. 15, 1969.

structure in the neighborhood of E_I are taken into account. Even if such information were available, it is doubtful that the additional insight gained would be worth the computational effort.

APPENDIX

The rate of diffusion to larger radii in velocity space is $v_r(l_v/r)$, where

$$v_r = \frac{\mathcal{E}e\tau_c/m}{\tau_c} = \frac{\mathcal{E}e}{m},$$

$$l_v = \frac{\mathcal{E}e\tau_c}{m},$$

$$r = v.$$

Thus,

$$\frac{dv}{dt} = v_r \frac{l_v}{r} = \frac{\mathcal{E}^2 e^2 \tau_c}{m^2 v},$$

But

$$\begin{aligned}\frac{dE}{dt} &= mv\frac{dv}{dt}\\ &= mv\frac{\mathcal{E}^2e^2\tau_c}{m^2v}\\ &= \frac{\mathcal{E}^2e^2\tau_c}{m} = \mathcal{E}^2e\mu = \mathcal{E}ev_d.\end{aligned}$$

PROCEEDINGS OF THE INTERNATIONAL CONFERENCE ON THE PHYSICS OF SEMICONDUCTORS, KYOTO, 1966
JOURNAL OF THE PHYSICAL SOCIETY OF JAPAN VOL. 21, SUPPLEMENT, 1966

X-8. Energy Losses by Hot Electrons in Solids: A Semiclassical Approach

A. ROSE

RCA Laboratories, Princeton, New Jersey, U.S.A.

The rates of energy loss by fast electrons to the various types of phonons and electronic excitations are derived in approximate form from an elementary classical model. Energy is emitted as a series of "energy-wells" in the wake of the moving electron. The quantum aspects are introduced as constraints on the classical model.

The rates of energy loss by fast electrons to the various types of phonons and electronic excitations have been analyzed and reported in the literature over the past fifty years. These analyses have for the most part been carried out in momentum space. The following is an attempt to restate the arguments in "real space" and in terms of an elementary model.

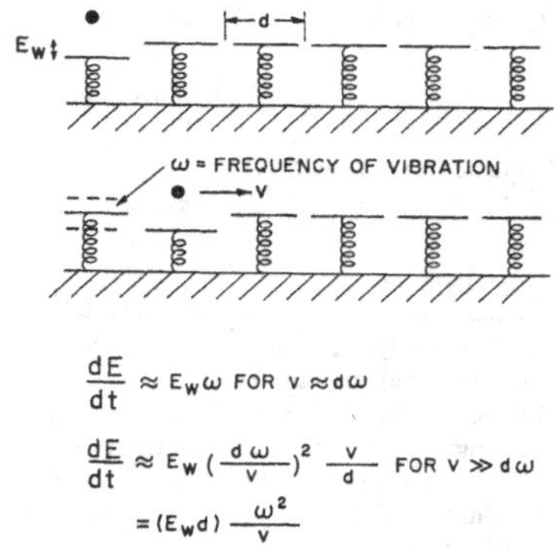

Fig. 1. Model for computing rates of energy loss.

The model in Fig. 1 shows a particle moving with velocity v past a series of elements, each of dimension d. The particle repels each element with a force such that, for a stationary particle, an energy E_w is stored in the compressed spring of the element. Also the frequency of vibration of each element is denoted by ω.

The maximum rate of loss of energy to the array is, by inspection,

$$\left.\frac{dE}{dt}\right|_{\max} \approx E_w \omega , \qquad (1)$$

and occurs at $v \approx \omega d$. Equation (1), applied to energy loss to polar optical phonons, gives $(\alpha\hbar\omega)\omega$ where $\alpha\hbar\omega$ is the interaction energy of the polaron. This agrees with the results of perturbation theory.[1,2]

The rate of loss of energy, in general, for $v > \omega d$ is

$$\frac{dE}{dt} = E_w \left(\frac{\omega d}{v}\right)^2 \frac{v}{d} = E_w d \frac{\omega^2}{v} . \qquad (2)$$

The factor $(\omega d/v)^2$ is the reduction in magnitude of the energy-well E_w due to transit of the particle past an element in times short compared with ω. The factor v/d is the number of wells traced out per second.

Equation (2) is the essence of the "real-space" argument. In order to apply it to electrons in a solid the value of the energy-well E_w must be determined in each case. E_w is the interaction energy between the stationary electron and a particular mode of energy loss in the medium.

For those cases where the electron couples to the medium via its coulomb field we can write:

$$E_w = \beta \frac{e^2}{Kd} ,$$

or, in differential form,

$$\Delta E_w = \beta \frac{e^2}{Kd^2} \Delta d ,$$

where β is, by definition, the fraction of the available coulomb energy $e^2/(Kd)$ that is used to form an energy-well. K is the dielectric constant for frequencies higher than that of the emitted radiation. Equation (3) is inserted into eq. (2) and integrated to give:

$$\frac{dE}{dt} = \beta \frac{e^2 \omega^2}{Kv} \ln \frac{mv^2}{\hbar\omega} , \qquad (4a)$$

in those cases where β and ω are constants. The limits of integration are the uncertainty radius $\hbar/mv$ and the adiabatic radius v/ω. In those cases where β or ω are not constant, the maximum

Table I. Rates of energy loss (dE/dt) by electrons of velocity v

Phenomenon	dE/dt	Source
Polar Optical	$\left[\frac{\varepsilon_0-\varepsilon_\infty}{\varepsilon_0}\right]\frac{e^2\omega^2}{\varepsilon_\infty v}\ln\left(\frac{2mv^2}{\hbar\omega}\right)$ Note: $1/2\,mv^2<\hbar\omega$	Frohlich[1] Callen[2]
Piezoelectric Phonons	$\frac{\pi}{4}\left[\frac{\varepsilon_p^2}{KC}\right]\frac{e^2\omega^2}{Kv}$ Note: $2\,mv=\hbar\omega/v_s$	Tsu[3]
Acoustic Phonons	$\frac{1}{4}\left[\frac{B^2\omega^2K}{4\pi e^2\rho v_s^4}\right]\frac{e^2\omega^2}{Kv}$ Note: $2mv=\hbar\omega/v_s$	Seitz[4] Conwell[5]
Non-Polar Optical Phonons	$\frac{1}{2}\left[\frac{\pi KD^2}{\rho e^2\omega^2\lambda^2}\right]\frac{e^2\omega^2}{Kv}$ Note: $1/2mv^2>\hbar\omega$ and $\frac{\lambda}{2\pi}=\frac{\hbar}{2mv}$	Conwell[5]
x-Ray Levels	$\left[\frac{\omega_p^2}{\omega_e^2}\right]\frac{e^2\omega_e^2}{v}\ln\frac{2mv^2}{\hbar\omega_e}$ Note: $\omega_p^2=\frac{4\pi ne^2}{m}<\omega_e^2$; $\hbar\omega_e$=Excitation energy of x-ray levels; $1/2mv^2>\hbar\omega_e$	Bohr[6] Bethe[7]
Plasma	$[1]\frac{e^2\omega^2}{v}\ln\left(\frac{2mv^2}{\hbar\omega}\right)$ Note: ω=plasma frequency and $1/2mv^2>\hbar\omega$	Bohm and Pines[8]
Cerenkov	$\left[1-\frac{1}{\varepsilon_\infty}\right]\frac{e^2\omega^2}{v}$ Note: $v=c$=velocity of light in vacuum	See e.g. Schiff[9]

ε_0 low frequency dielectric constant
ε_∞ high frequency (optical) dielectric constant
K dielectric constant
ε_p piezoelectric constant
C elastic modulus (dynes/cm²)
ρ density (grams/cm³)
v_s phase velocity of sound
v velocity of electron
ω angular frequency of radiation
m effective mass of electrons
B deformation potential (electron volts in ergs/unit strain)
D optical deformation potential (electron volts in ergs per centimeter relative shift of sublattices)

contribution to the integral occurs in the neighborhood of a particular radius and becomes:

$$\frac{dE}{dt}\approx\beta\frac{e^2\omega^2}{Kv}. \tag{4b}$$

Equation (4) is the form in which the energy-loss expressions in Table I are recast from the literature. The values of β are set off in square brackets.

It can be shown by an elementary argument that for the losses to phonons β is also equal to the ratio of electrical to total energy of the corresponding macroscopic sound waves. [See RCA Rev. **27** (1966) 96 for calculations of β and their relation to the acoustoelectric effects]. For acoustic and optical phonons coupled by deformation potentials, β was computed as if the slopes of the deformed band edges were a macroscopic electric field. The same result is also obtained directly from eq. (2) by computing the energy well E_w formed by an electron with an uncertainty radius $\hbar/(mv)$.

The value of β for plasmons is unity since the coulomb field of the "stationary" electron is substantially completely cancelled by the polarization field induced in the plasma. Similarly, in the case of the deep-lying or x-ray levels, only the fraction ω_p^2/ω_e^2 of the coulomb field of the electron is cancelled by the induced polarization field. Hence, it is only this fraction of the coulomb energy that is used in forming an energy well.

The quantum constraints on the classical model appear as the threshold conditions that the electron must have at least the energy and momentum of the quanta or radiation it emits. Also, the radius of the electron within which its charge is effectively smeared out is given by the uncertainty relation $r \approx \hbar/(mv)$.

This paper is based on work done in collaboration with Professor George Whitfield, Penn State University. The work was supported in part by the U.S. Army Research Office-Durham, Durham North Carolina and by the U.S. Army Engineering Research and Development Laboratory, Fort Belvoir, Virginia.

References

1) H. Frohlich: Proc. Roy. Soc. (London) **A188** (1947) 521.
2) H. B. Callen: Phys. Rev. **76** (1949) 1394.
3) R. Tsu: J. appl. Phys. **35** (1964) 125.
4) F. Seitz: Phys. Rev. **76** (1949) 1376.
5) E. M. Conwell: Phys. Rev. **135** (1964) A1138.
6) N. Bohr: Phil. Mag. **25** (1913) 10; **30** (1915) 581. (see also Jackson for critical historical review).
7) H. Bethe: Ann. Phys. **16** (1933) 285; *Handbuch der Physik* ed. Geiger and Scheel (Springer, Berlin, 1933) Vol. 24, Part I. (see also Jackson for critical historical review).
8) D. Bohm and D. Pines: Phys. Rev. **92** (1953) 609; L. Marton *et al.*: *Advances in Electronics*, ed. L. Marton (Academic Press, N. Y., 1955) Vol. 7, p. 230.
9) L. I. Schiff: *Quantum Mechanics* (McGraw Hill Book Co., New York, 1955) p. 271.
10) J. D. Jackson: *Classical Electrodynamics* (John Wiley and Sons, New York 1963).

DISCUSSION

Landsberg, P. T.: Could you say if loss by impact ionization could be included in your last Figure ?

Rose, A.: The application of this model to impact ionization is included in the complete paper due to appear shortly in the RCA Review.

Reprint from COOPERATIVE PHENOMENA

Edited by H. Haken and M. Wagner

Springer-Verlag Berlin Heidelberg New York 1973. Printed in Germany. Not in trade

Classical Aspects of Spontaneous Emission

ALBERT ROSE

RCA Laboratories, Princeton, N. J./U.S.A.

Abstract

Spontaneous emission is frequently ascribed to the action of zero point vibrations on an excited electron. If this were so, there would be no classical parallel since classical media do not have zero point vibrations. In contrast to the model using zero point vibrations, it is shown that spontaneous emission by a moving electron in a solid can be accounted for in terms of a classical interaction between electron and solid. While the mechanism of the interaction is classical, the evaluation of its magnitude is subject to certain quantum constraints.

1. Introduction

An excited electron relaxes towards lower energy states at a rate, according to the quantum mechanical formalism, proportional to $n+1$ where n is the density of quanta per mode in the field and the term unity is ascribed to "spontaneous emission". The action of the n quanta in the field, both in exciting the electron to higher energy states and in stimulating the emission of energy by an electron, has a natural classical parallel in the action of periodic forces on resonant systems. These forces can, depending on their phase, either excite or damp the vibrations of the resonant system. This leaves the question of whether "spontaneous emission" also has a classical equivalent. The answer should be "obviously, yes" as we will point out below. The literature, however, is far from unanimous in supporting this answer.

Several examples of the interpretations of spontaneous emission are worth citing here. At the outset, the term "spontaneous" was BOHR's translation of EINSTEIN's phrase "emission without excitation by external causes" [1]. Either description was certain to endow spontaneous emission with a certain mystery. Nor does the formalism of quantum mechanics offer any relief. Spontaneous emission is simply one of the consequences of this formalism. Quantum mechanics ignores any questions of "what causes it". Indeed, any answer to the question would belie the name "spontaneous".

A number of authors associate spontaneous emission with the presence of zero point vibrations. The following quotation is from SCHIFF [2] "From a formal point of view, we can say that the spontaneous emission probability is equal to the probability of emission that would be induced by the presence of one quantum in each state of the electromagnetic field. Now we have already seen that the smallest possible energy of the field corresponds to the presence of one-half quantum per state. This suggests that we regard the spontaneous emission as being induced by the zero-point oscillations of the electromagnetic field; note, however, that these oscillations are twice as effective in producing emissive transitions as are real photons and are of course incapable of producing absorptive transitions."

PARK and EPSTEIN [3] derive the Planck distribution by treating the energy of the zero point vibrations on a par with real quanta in their effect on stimulating emission. They avoid the factor of two discrepancy by assigning $1/2\, h\nu$ to each of the two "degrees of freedon" of the zero point vibrations. If the degrees of freedom are taken to mean kinetic and potential energy, then each mode of zero point vibrations will have an energy $h\nu$ rather than the $1/2\, h\nu$ that is usually assigned to the ground state. A recent tutorial paper by SCULLY and SARGENT [4] on "The Concept of the Photon" contains the phrase "... that is, the vacuum fluctuations 'stimulate' the atom to emit spontaneously". The quotes on stimulate are theirs.

The concept that spontaneous emission is caused by the action of zero point vibrations on the excited state has a simple logical consequence. Since classical physics does not contain zero point vibrations, classical physics can not account for spontaneous emission. Spontaneous emission must then be one of the novelties introduced by quantum mechanics and must be completely outside the province of classical physics.

In contrast to these conclusions, we believe that classical physics does provide the physical mechanism for spontaneous emission. This is simply the interaction of an electron with its surrounding medium. Moreover, a semi-classical analysis of this interaction yields the correct average rate of radiation of energy by the electron to its surround. The term semi-classical means here that the classical interaction between electron and medium is retained *as the mechanism or physical cause for spontaneous emission* but that the size of the electron is given by its quantum mechanical uncertainty radius, $\hbar/mv$, where m is, in the case of solids, the effective mass of the electron.

While classical physics provides the physical mechanism for spontaneous emission, and a semi-classical analysis gives the average rate of emission, the actual detailed emission in the form of quanta of energy radiated stochastically in time and space is purely a quantum mechanical phenomenon and has no generally agreed classical source. Stated in other terms, causality, in the deterministic sense, is preserved classically in the average and violated (or not understood) in the fine. This is different from stating that the phenomenon of spontaneous emission is completely outside the sphere of classical physics.

2. Spontaneous Emission in Solids

The quotations from the literature, which we cited, all had in mind the effect of zero-point vibrations in vacuum on causing the spontaneous emission of photons by excited electrons. The quantum mechanical formalism for the spontaneous emission of phonons by energetic electrons in solids completely parallels that for

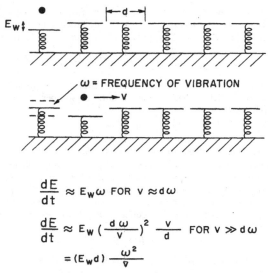

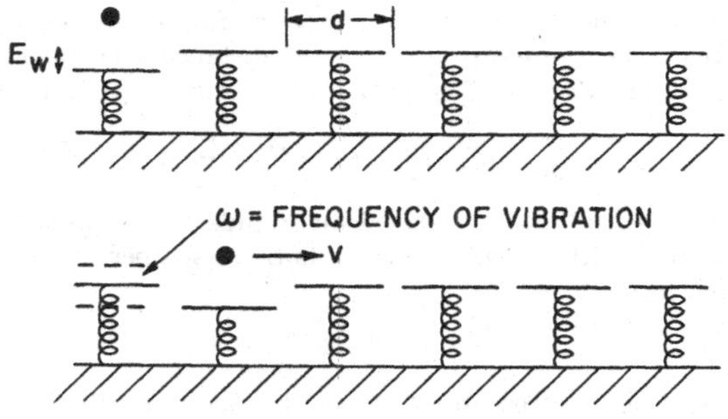

$$\frac{dE}{dt} \approx E_W \omega \quad \text{FOR } v \approx d\omega$$

$$\frac{dE}{dt} \approx E_W \left(\frac{d\omega}{v}\right)^2 \frac{v}{d} \quad \text{FOR } v \gg d\omega$$

$$= (E_W d)\,\frac{\omega^2}{v}$$

Fig. 1. Classical model for spontaneous emission by a moving electron in a solid

the emission of photons in vacuum. The rate of emission in solids is again proportional to $n+1$ where n is the modal density of phonons. The term unity is ascribed to spontaneous emission and is equally susceptible, as in vacuum, of being ascribed to the presence of the zero point vibrations of the phonon states.

We outline here the classical mechanism for spontaneous emission by "hot" electrons in solids. The choice of a solid rather than a vacuum medium is deliberate because the interaction of the electron with its surround is obvious and palpable. At the same time it would be artificial to expect that spontaneous emission in solids would be fundamentally different from that in vacuum in the sense that the classical interaction with the medium causes one and the non-classical zero-point oscillations cause the other.

The accompanying figure shows schematically the classical model for spontaneous emission in solids [5]. A particle moves past an array of elements with a velocity v. Each element has a dimension d and is attached to a spring such that it has a characteristic frequency ω. There is a force of repulsion between the particle and each element such that, when the particle is stationary, an energy E_w is stored in the element opposite the particle. In this model, the particle represents the electron, the array of elements represents the medium, and the energy E_w is the interaction energy between electron and medium.

In effect, an electron polarizes (or, in other ways, deforms), the solid medium. In so doing, it imparts an energy E_w to the medium. A slow moving electron carries this polarization energy with it as in FRÖHLICH's [6] model of the polaron. When the electron moves with a velocity $v \approx d\omega$ it succeeds both in almost completely polarizing the medium and at the same time in leaving most of this polarization energy behind. It then achieves a maximum rate of energy loss $E_w \omega$. At still higher velocities, the electron passes the distance d in a time too fast to completely polarize the medium. The polarization energy left in its trail is then

$$\frac{dE}{dt} = E_w \left(\frac{d\omega}{v}\right)^2 \frac{v}{d} = E_w \frac{d\omega^2}{v} \tag{1}$$

where the factor $E_w (\omega d/v)^2$ is the reduced magnitude of the polarization energy and v/d is the number of the polarized elements traversed per second. The energy E_w can be written formally as

$$E_w = \beta \frac{e^2}{Kd} \tag{2}$$

where β is the fraction of coulomb energy of the electron, e^2/Kd, which can be imparted to the medium. K is the electronic part of the dielectric constant. β depends on the physical mechanism by which the electron couples to the medium. Its various forms include polarization by ionic displacements, the piezoelectric effect and the deformation of the energy band structure. Insertion of Eq. (2) in Eq. (1) yields the familiar general form for rates of energy loss:

$$\frac{dE}{dt} = \beta \frac{e^2 \omega^2}{Kv}. \tag{3}$$

Eq. (3) does not include a geometric factor arising from integrating over a range of values of d. It is, however, sufficient to bring out the classical physics which underlies the spontaneous emission of phonons by energetic electrons in solids.

The argument thus far is purely classical and in no sense does it make use of the existance of zero point vibrations.

The evaluation of Eq. (3) does, however, make use of quantum mechanics, but only as a constraint on the emission and not as a cause of it. The smallest value for d, for example, is given by the uncertainty radius $\hbar/mv$ of the electron. Further, m is the effective mass of the electron in the solid. This is a quantum mechanical concept. So, also, is the evaluation of β for the case of deformation of the energy band structure. Finally, the electron must have at least the crystal momentum and the energy of the phonon it emits. The classical argument for Eq. (3) was first given by Fröhlich and Pelzer [7] for ionic solids in 1948.

We emphasize, again, that the physical cause for spontaneous emission in a solid is the classical interaction of the electron with its surrounding medium. Quantum mechanics enters in as a constraint on the evaluation of this interaction and not as the cause of the spontaneous emission itself.

3. Discussion

It is clear in the case of an electron interacting with a solid medium that the coupling constant β must be symmetric*. β is a measure not only of the extent to which the electron deforms its medium but also, at the same time, the extent to which the deformed medium acts on the electron. This symmetry has been explored in detail in Ref. [5]. The action of the deformed medium on the electron is the essence of the acousto-electric effects. The action of the electron

* The symmetry of coupling constants is a general requirement of detailed balance.

on the medium to deform it is the essence of spontaneous emission. Both are characterized by the same coupling constant β. This symmetry is already well known in the case of the piezoelectric effect where a deformed crystal produces an electric field and, at the same time, an external electric field is able to deform the crystal. Both are characterized by the same electromechanical coupling constant.

We explore now the nature of the deformation that an electron effects on its medium. We recognize that, by virtue of the $(n+1)$ formalism, a set of pre-existing zero point oscillations, each having an energy of one quantum, is able *formally* to account for spontaneous emission. But if we ascribe spontaneous emission to the deformation of the medium by the presence of the electron, the deformation has no existence in the absence of the electron. *The insertion of the electron causes the deformation whose action on the electron is equivalent to the action of a pre-existing set of zero point oscillations.* The zero point oscillations constitute, then, a formal representation of the interaction between electron and medium.

This discussion has centered on the emission of phonons by energetic electrons in solids. It is shown in Ref. [5] that the emission of photons via Cerenkov radiation also follows the same formalism. In this case, the electron deforms the electronic part of the solid rather than displacing ionic cores.

One consequence of regarding the zero-point oscillations as a description of the action of the electron on its medium is that there is a natural cut off towards short wavelengths given by the size of the electron. In a solid, this cut off is the uncertainty radius $\hbar/mv$ of the electron. In vacuum, the same assignment should be valid. In any event, the infinities that arise from the use of a pre-existing set of zero point oscillations with no natural cut off do not arise here. The parallel in vacuum of the polarization energy in the solid is the coulomb energy of the electron.

In summary, the present discussion of spontaneous emission in solids has attempted to remove its spontaneity and to assign its cause to the classical polarization (or deformation) of the medium by the electron. Quantum mechanics enters in as a constraint on the evaluation of the magnitude of the classical interaction between electron and medium.

References

1. See e.g. The Conceptual Development of Quantum Mechanics by M. Jammer. New York: McGraw-Hill Book Co. 1966.
2. Schiff, L. I.: Quantum Mechanics, 3rd Ed., p. 533. New York: McGraw-Hill Book Co. 1968.
3. Park, D., Epstein, H. T.: On the Planck Radiation Formula. Am. J. Phys. **17**, 301 (1949).
4. Scully, Marlan O., Sargent III, Murray: The Concept of the Photon. Physics Today, p. 46, March (1972).
5. This classical mechanism is used to derive the spectrum of energy losses by energetic electrons in solids in the series of papers: The Acousto-electric Effects and the Energy Losses by Hot Electrons, Rose, A.: RCA Rev. **27**, 98 (1966); **27**, 600 (1966); **28**, 634 (1967); **30**, 435 (1969); **32**, 463 (1971).
6. Fröhlich, H., Pelzer, H., Zienan, S.: Properties of Slow Electrons in Polar Materials. Phil. Mag. **41**, 221 (1950).
7. Fröhlich, H., Pelzer, H.: Polarization of Dielectrics by Slow Particles. Reports of the British Electrical and Allied Industries Research Association #L/T 184 (1948).

phys. stat. sol. (a) 61, 133 (1980)

Subject classification: 20; 6; 13.5

Corporate Research Science Laboratories, Exxon Research & Engineering Co., Linden[1])

Spontaneous Emission Revisited

By

A. Rose

Dedicated to Dr. W. J. Merz on the occasion of his 60th birthday

The cause of spontaneous emission is frequently assigned to the presence of zero-point energy in the radiation field. As such it would have no classical equivalent since zero-point energy does not exist classically. It is argued that spontaneous emission is not caused by zero-point energy but rather by an essentially classical interaction between the excited electron and its surrounding medium.

Die Ursache für die spontane Emission wird häufig in der Nullpunktsenergie des Strahlungsfeldes gesucht. Als solche hätte sie nichts Entsprechendes in der klassischen Physik, da es dort keine Nullpunktsenergie gibt. Es wird gezeigt, daß die spontane Emission nicht durch die Nullpunktsenergie hervorgerufen wird, sondern vielmehr durch eine ihrem Wesen nach klassische Wechselwirkung zwischen dem angeregten Elektron und dem umgebenden Medium.

1. Introduction

In the course of the last several years I have made an informal canvass of my colleagues with the question: "Given an excited hydrogen atom isolated in a zero-temperature space, what causes it to radiate?" After some hesitation the reaction has usually been that the hydrogen atom is in some way "tickled" or persuaded by the zero-point energy in space to radiate. This reaction reflects the treatment of "spontaneous emission" in the literature. For the most part, the question is not even raised in the literature. Instead, the formalism of QED (quantum electrodynamics) is invoked as the valid means of computing the rate of spontaneous emission. Occasionally, an author is impelled to raise the question to himself and to reply, as did my colleagues, that the mechanism responsible for spontaneous emission is the universal presence of zero-point energy in space.

Thus, for example, Eisberg and Resnick [1]: "Like any other system with discretely quantized energy, the electromagnetic field has a zero-point energy. Thus quantum electrodynamics shows that there will always be some electromagnetic field vibrations present, of whatever frequency is required to induce the charge oscillations that cause the atom to radiate "spontaneously". Schiff [2] notes "... This suggests that we regard spontaneous emission as being induced by the zero-point oscillations of the electromagnetic field; note, however, that these oscillations are twice as effective in producing emissive transitions as are actual quanta ...". The last remark stems from the fact that spontaneous emission occurs at a rate that would be induced by the presence of one quantum per degree of freedom whereas the zero-point oscillations provide only one half quantum. Again, Baym [3] writes: "In a sense, one can regard spontaneous emission as induced emission due to the vacuum fluctuations of the electromagnetic field".

---

[1]) Linden, New Jersey 07036, USA.

These three references are cited only to show that our initial question is not an empty one and also to provide the backdrop for the thesis of this discussion that zero-point oscillations play no role in spontaneous emission. In fact, our thesis is that "spontaneous emission" is a misnomer. Spontaneous emission is no more spontaneous than the excitation of a hydrogen atom from the ground state by the presence of a density of one "real" quantum per degree of freedom, of the appropriate energy, in the radiation field. Stated in other terms, the emission of the one quantum of excitation by the hydrogen atom into an empty radiation field is the symmetric counterpart of the emission of one quantum from the radiation field into the hydrogen atom, initially in its ground state, when the density in the radiation field is one quantum per degree of freedom. In both cases one has a system A coupled to a system B such that excitation energy introduced into either system will be transferred by the coupling to the other system. The corollary to this statement is that the zero-point energy is not the coupling mechanism. Rather, the coupling is the classical disturbance provoked by the presence of excess energy in one system on the second system.

We will carry out the argument in terms of the emission of phonons by excited electrons in a semiconductor because the sense of our statements is more readily apparent there. We will return to the hydrogen atom at the end. It will be recognized that the formal aspects of the emission of photons and phonons are the same as far as the nature of spontaneous emission is concerned. We will introduce these arguments by first examining the overriding implications of detailed balance.

The "revisited" in the title of this discussion stems from a series of papers [4] in which electron transport in solids was treated by first computing the rate of spontaneous emission of phonons by a classical argument on which certain quantum constraints were imposed. This procedure has the virtues of presenting the electron–phonon interactions in configurational (rather than Fourier) space, of offering at least a semiquantitative understanding of high mobilities in amorphous solids [5], and of retaining the same formalism as the electron–phonon interaction progresses from weak to strong.

2. Detailed Balance

Einstein [6] applied an argument of detailed balance between atoms immersed in a sea of radiation to obtain Planck's law and the relation between the coefficients A and B characterizing the rates of spontaneous and induced emission, respectively. His only apparently non-classical assumption was that of discrete states of excitation for the atom [7].[2])

The form of Einstein's result that we wish to emphasize here is:

$$\text{rate of absorption of radiation by an atom} = \beta n\,, \tag{1}$$

$$\text{rate of emission of radiation by an excited atom} = \beta(n+1)\,. \tag{2}$$

Here n is the Planck density of photons per mode,

$$n = \left(\exp\frac{\hbar\omega}{kT} - 1\right)^{-1}, \tag{3}$$

and β is a coupling constant between the radiation and the atom. Equations (1) and (2) state that the coupling constant is the same for radiation-induced emission and

[2]) In this provocative and instructive analysis of Einstein's paper, Lewis calls attention to the fact that Einstein's assumption of discrete states was shown not be necessary to the argument. He states: "The key point is not the assumption of discrete states but rather that the atom is almost always in some particular energy state and spends negligible time in changing its state".

absorption, on the one hand, and for spontaneous emission by the excited atom on the other hand. The terms n in (1) and (2), of course, refer to radiation-induced emission and absorption while the term unity in (2) refers to spontaneous emission. It is this term unity which, as we cited in the introduction, is usually assigned to the effect of zero-point energy in the radiation field on the excited atom: namely, the zero-point energy induces the atom to emit.

Note at this point that the term unity was obtained by an argument of detailed balance and was not dependent on mechanism. That is, whatever mechanism one introduces to account for spontaneous emission must yield the term unity in (2).

For example, the formalism of quantum electrodynamics yields the same term unity by virtue of its particular choice of commutation relations. To quote Schiff again: "... the spontaneous emission arises from the 1 in the factor $(n + 1)$ ... This in turn comes from the commutation relations and *hence is a purely quantum-mechanical effect*".

The underlining is mine. It calls attention to what I believe is an inversion of priorities. In contrast to Schiff, I would say that the mechanism-free argument of detailed balance is the dominant and overriding source of the term unity and that any particular mechanism such as the quantum-mechanical formalism of QED had to match the result obtained from detailed balance. This, for example, does not preclude the interpretation of the term unity in terms of some other purely classical mechanism. Indeed, if any example of classically provoked spontaneous emission can be identified, it should cast doubt on Schiff's statement that spontaneous emission is a purely quantum-mechanical effect. We will return to this subject in the discussion.

At this point we present another detailed balance argument which appears to cast doubt on the spontaneity of spontaneous emission.

Consider two systems, A and B, in thermal equilibrium. The two systems are in contact and therefore exchange energy. It must be true that

$$\left.\frac{\mathrm{d}E}{\mathrm{d}t}\right|_{\mathrm{A}\to\mathrm{B}} = \left.\frac{\mathrm{d}E}{\mathrm{d}t}\right|_{\mathrm{B}\to\mathrm{A}} . \tag{4}$$

We take, as a particular embodiment, an atom for system A in contact with a radiation field for system B. Further, let the energy of the first excited state of the atom be such that $E/kT \gg 1$, then

$$n = \left[\exp\left(\frac{E}{kT}\right) - 1\right]^{-1} \cdot \exp\left(\frac{-E}{kT}\right) \tag{5}$$

and the rates of energy flow between A and B, following (1) and (2), are

$$\beta \exp\left(\frac{-E}{kT}\right)\left[\exp\left(\frac{-E}{kT}\right) + 1\right] \rightleftarrows \beta \exp\left(\frac{-E}{kT}\right). \tag{6}$$

Equation (6) reduces to

$$\beta \exp\left(\frac{-E}{kT}\right) \rightleftarrows \beta \exp\left(\frac{-E}{kT}\right) \tag{7}$$

for $E/kT \gg 1$.

The left-hand side of (7) states that during the fraction of time $\exp(-E/kT)$ that the atom is in the excited state, it radiates, by spontaneous emission, at the rate β. The right-hand side of (7) states that, owing to the fractional occupancy $\exp(-E/kT)$

of modes in the radiation field, having an energy $\hbar\omega = E$, the rate of radiation-induced absorption in the atom is $\exp(-E/kT)\,\beta$. All of the above may appear obvious and trivial. It is spelled out here in order to emphasize that the physics of spontaneous emission contained in the factor β is the same as that for induced absorption which depends on the same factor β. The coupling constant β represents, on the one hand, the interaction between the atom and the field of radiation and, on the other hand, the interaction between the field of radiation and the atom. It is the same physical coupling viewed from either end. *Hence, spontaneous emission is no more spontaneous than the excitation of the atom by the presence of quanta ($\hbar\omega = E$) in the radiation field.*

Furthermore, if it can be shown (as in the following section) that spontaneous emission is predominantly a classical phenomenon, it must then follow that the zero-point energy in the radiation field can play no role since zero-point energy is not present in the classical world.

3. Spontaneous Emission in Solids

Consider an electron in the conduction band of a solid and in thermal equilibrium with the phonons in the solid. The exchange of energy between the electron and the phonon field follows the same formalism of (1) and (2) as does the exchange of energy between an excited atom and a radiation field. In the solid, however, one can identify the physics of spontaneous emission, and therefore the coupling constant β, as predominantly classical.

A generalized argument for spontaneous emission in a solid was obtained [4] from the interaction of the Coulomb field of the electron with the solid. The electron polarizes the solid and gives rise to a polarization energy in the solid in the neighborhood of the electron. If the electron moves with a velocity less than that of the phonons, it carries this polarization field with it. If its velocity exceeds that of the phonons, it *must* leave behind it the energy of polarization which is then radiated as phonons. This will be recognized as a type of Čerenkov radiation equally valid for the emission of phonons as well as photons and for the utterly classical radiation of water waves in the wake of a moving boat.

We are not interested here in carrying out the details of the argument for the spontaneous emission of phonons. The details are contained in [4] and include the rates of emission of energy to polar and non-polar optical phonons, acoustic phonons coupled by deformation potentials and by the piezoelectric effect, plasmons, X-ray excitations, and Čerenkov photons. The well-known expressions derived from perturbation theory were matched by the predominantly classical polarization argument.

We are interested here in certain aspects of this predominantly classical argument that have a bearing on the question of what causes spontaneous emission. The term "predominantly classical" means that the argument for the rate of loss of energy via the trail of polarization is a classical argument. The quantitative evaluation of the rate of loss of energy is, however, subject to the non-classical constraint that the electron is not a point charge but has an uncertainty radius

$$r = \frac{\hbar}{m^* v} \tag{8}$$

and cannot polarize the solid within this radius. m^* is the effective mass of the electron and is a consequence of the non-classical wave nature of the electron.

Also, the quantization of phonons means that the classically continuous rate of energy loss $\mathrm{d}E/\mathrm{d}t$ computed for the electron must be equated to the quantized and stochastic rate $\hbar\omega/\tau$ where τ is the mean time to emit a phonon:

$$\left.\frac{\mathrm{d}E}{\mathrm{d}t}\right|_{\text{classical}} = \frac{\hbar\omega}{\tau}. \tag{9}$$

These constraints, then, take into account the wave nature of the electron and the quantized nature of the phonons. They are imposed on the classical polarization argument. That is, the argument and the mechanism for spontaneous emission is classical; its quantitative evaluation and interpretation is subject to certain quantum-mechanical constraints.

The general expression in [4] for the rate of loss of energy by the electron to phonons was cast in the form

$$\frac{\mathrm{d}E}{\mathrm{d}t} \approx \beta\frac{e^2\omega^2}{K_{\mathrm{H}}v}, \tag{10}$$

where ω is the frequency of the emitted phonons, v the velocity of the electron, K_{H} the electronic part of the dielectric constant, and β the coupling constant. A numerical factor of order unity is omitted from (10) and arises from integrating the polarization energy over the range of radii from the uncertainty radius $\hbar/(m^*v)$ to the outer limit $v/r = \omega$ beyond which the solid is reversibly polarized by the passing electron. The coupling constant β has the interesting, general form

$$\beta = \left.\frac{\text{electrical energy}}{\text{total energy}}\right|_{\text{lattice deformation}}. \tag{11}$$

Its meaning is that the energy necessary to effect a distortion of a solid can be divided into an elastic and an electric field component. It is the electric field component that couples the phonon field to the electron and vice versa. Hence, the ratio of the electric field energy to the total energy (sum of electric field energy and elastic energy) is the measure of the interaction.

This form of the coupling constant has, of course, a clearly classical flavor. The same coupling constant was found to be valid not only for the quantum process of spontaneous emission of phonons but also for the purely classical phenomena of the acousto-electric effect where the absorption of sound waves by free electrons causes an electric current or conversely, where an electric current can amplify the sound wave when the electrons drift faster than the speed of sound.

It is instructive to look at this coupling constant for the case of an electron emitting acoustic phonons coupled by the piezoelectric effect. On the one hand, the coupling constant gives the fraction of the Coulomb field energy of the electron which is converted into an elastic deformation of the surrounding solid. It is the latter energy that is radiated by "spontaneous" emission as phonons. On the other hand, the coupling constant also gives the fraction of the phonon (or acoustic wave) energy that can be transmitted to the electrons in the process of induced absorption. We call attention here to the fact that even though the coupling constant is the same for spontaneous emission and induced absorption (or emission), it looks somewhat different when viewed from the points of view of the electron acting on the medium and the medium acting on the electron.

The parallel to the above in the case of the hydrogen atom is that the Coulomb energy of the electron is what couples the electron to the radiation field in spontaneous emission; the electric field energy of the photon is what couples the light wave to the electron in induced absorption (or emission). In this case, the coupling constant β is unity since all of the energy in the distorted medium (vacuum) is electrical. The major difference between the spontaneous emission of the hydrogen atom and that of the electron in a solid is that the electron in vacuum cannot move faster than the speed of light in order to disengage itself from its Coulomb field energy. Rather, the electron must now execute an accelerated motion in order to disengage itself from its Coulomb field energy. This disengagement is not as effective as the Čerenkov process in a solid so that only a small fraction $(e^2/\hbar c)^2$ of the Coulomb energy is shaken loose per oscillation of the electron. What we have described here is, of course, only the classical process of radiation from an accelerated electron.

4. Discussion

The mechanism-free argument of detailed balance leads to the relations:

$$\text{rate of absorption of radiation by an atom} = \beta n\,, \tag{1}$$

$$\text{rate of emission of radiation by an excited atom} = \beta(n+1)\,. \tag{2}$$

Customarily, the coupling constant β is obtained from the matrix element connecting two states of the electron via the perturbation by the electromagnetic field of the radiation. In brief, the matrix element is obtained via the formalism of quantum mechanics applied to induced emission and absorption. This is sufficient, of course, to obtain the coupling constant for spontaneous emission. However, when it comes to interpreting spontaneous emission, in particular, to assigning a cause to it, the literature is somewhat at a loss since the electric fields of the radiation field that provoked the induced emission and absorption are now absent. The electron now radiates into an apparently empty medium spontaneously, that is, in the words of Einstein "without any external cause". In order to avoid such an abhorrent intellectual vacuum, an event without a cause, it appears that a number of authors have siezed upon the ever present zero-point energies in the radiation field as the "cause" of the spontaneous emission. Even though the density of zero-point energy is $\frac{1}{2}$ quanta per mode and the argument of detailed balance calls for the equivalent of one quantum per mode, the authors nevertheless feel that they have rescued spontaneous emission from its otherwise mysterious implications.

We have chosen here to approach the problem of spontaneous emission directly. In the case of spontaneous emission of phonons by an excited electron in a solid we have shown that the spontaneous emission is due to a Čerenkov process whereby the polarization energy induced by the Coulomb field of the electron in the surrounding medium is left behind in the wake of the moving electron to be radiated as phonons. This examination has shown spontaneous emission to be an essentially classical phenomenon which does not depend on or make use of the zero-point energy of the phonon field. Rather, spontaneous emission is no more spontaneous than its counterpart, the excitation of an electron by the presence of phonons.

Since the detailed balance relations (see (1) and (2)) apply equally to electrons exchanging energy with a phonon field and atoms exchanging energy with a photon field, it would be a surprising turn of logic to find that the classical phenomenon of polarization of the medium by the electron is the cause of spontaneous emission in the one case and the non-classical phenomenon of zero-point energy is the cause in the second case. It is at least logically more economical to look for the same mechanism in both cases, namely, the ability of the moving electron to disengage itself from its Coulomb-field-induced energy in the surrounding medium. This leads to the classical Čerenkov type of spontaneous radiation in the solid and to the classical accelerated electron type of spontaneous radiation in vacuum. Indeed, all three authors [1 to 3] compute the spontaneous emission for the hydrogen atom using the classical accelerated electron argument and show that the result matches that obtained by QED. At the same time, they cite the zero-point energy as the cause of spontaneous emission. What they do not do is to attempt to clarify why these two completely unrelated causes can lead to the same result. Our thesis here is that their interpretation of spontaneous emission in terms of zero-point energy is a gratuitous afterthought. The zero-point energies were not initially inserted into their QED calculations but were brought in afterwards to paper over an otherwise puzzling phenomenon.

The zero-point energies are related to the mechanism for spontaneous emission by virtue of a common parentage. They are not related as cause and effect. The common parentage is the fact that an electron interacts with its surrounding medium. The logic of this relationship can be sketched in the form

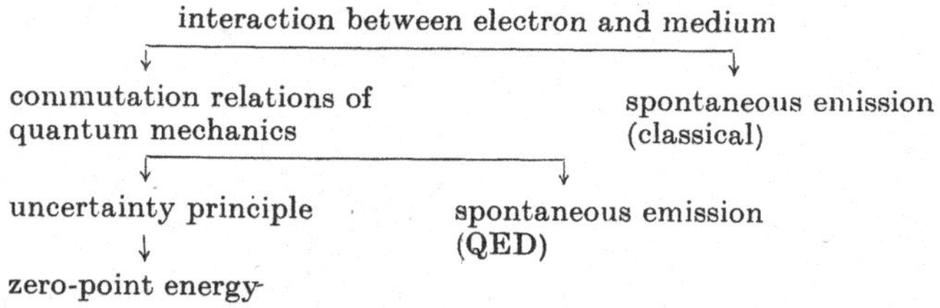

The interaction between an electron and its medium via its Coulomb field is obvious from the classical approach. In quantum mechanics, this interaction is introduced via the formalism of commutation relations. The commutation relations, in turn, yield the uncertainty principle which is the source of the zero-point energy. The commutation relations of QED also yield (see e.g. Schiff) the term describing spontaneous emission, as they must, in order to satisfy detailed balance. Hence, zero-point energy and spontaneous emission are both consequences of the interaction between an electron and its medium. One is not the cause of the other. In the classical approach, spontaneous emission is a direct and unambiguous consequence of the interaction between an electron and its medium.

One consequence of the classically rooted calculation of spontaneous emission described in this paper has not been emphasized and deserves at least a passing mention. The classical analysis of spontaneous emission means that the coupling constant β is essentially classical. Hence, since the same coupling constant holds for induced emission and absorption, one should expect that these processes should also yield to a predominantly classical approach.

We conclude with a reference to a recent paper [8] in which spontaneous emission is shown to be a consequence of an intrinsic instability of the excited state, owing to the self-interaction of the electron, and not to involve zero-point energy. The bridge between the predominantly classical concepts of the present paper and the more formal quantum-mechanical analysis of DeGregoria has yet to be made.

References

[1] R. Eisberg and R. Resnick, Quantum Physics, John Wiley, Inc., New York 1974 (p. 316).
[2] L. I. Schiff, Quantum Mechanics, 3rd ed., McGraw-Hill Publ. Co., Inc., New York 1955 (p. 533).
[3] G. Baym, Lectures on Quantum Mechanics, W. A. Benjamin, Inc., Reading (Mass.) 1973 (p. 276).

[4] A. Rose, RCA Rev. **27**, 98, 600 (1966); **28**, 634 (1967); **30**, 435 (1969); **32**, 463 (1971).
[5] W. F. Schmidt, Photoconductivity and Related Phenomena, Ed. J. Mort and D. M. Pai, Elsevier Publ. Co., New York 1976 (p. 335).
[6] A. Einstein, Phys. Z. **18**, 121 (1917).
[7] H. R. Lewis, Amer. J. Phys. **41**, 38 (1973).
[8] A. J. DeGregoria, Nuovo Cimento **51A**, 377 (1979).

(Received July 10, 1980)

From: PHYSICS OF DISORDERED MATERIALS
Edited by David Adler, Hellmut Fritzsche
and Stanford R. Ovshinsky
(Plenum Publishing Corporation, 1985)

A SIMPLE CLASSICAL APPROACH TO MOBILITY IN AMORPHOUS MATERIALS

Albert Rose

Visiting Scientist, Exxon Research and Engineering Co.
Clinton, NJ 08801

ABSTRACT

A simple and essentially classical formalism, valid for both amorphous and crystalline materials, is used to compute the phonon component of electron mobilities in amorphous materials. A significant part of this formalism is a simple, physical concept of a coupling constant valid for all of the electron interactions in a solid both classical and quantum.

SOME BACKGROUND COMMENTS

Over the years I have had occasion to confirm a long persistant dichotomy in the approach of physicists to the various phenomena of "spontaneous emission". In casual lunch table conversations wih many of my colleagues I have raised the question: "Given an excited hydrogen atom at zero degrees Kelvin and removed from any external disturbances, what causes it to spontaneously emit a photon?" Almost without exception the cause of the emission has been ascribed to the presence of zero-point quanta. This reply reflects the more considered quotations from well known text books. Thus, for example, Eisberg and Resnick[1]: "Like any other system with discretely quantized energy, the electromagnetic field has a zero-point energy. Thus, quantum-electrodynamics shows that there will always be some electromagnetic field vibrations present, of whatever frequency is required to induce the charge oscillations that cause the atom to radiate "spontaneously". Schiff[2] notes "... This suggests that we regard spontaneous emission as being induced by the zero-point oscillations of the electromagnetic field; note, however, that these oscillations are twice as effective in producing emissive transitions as are actual quanta ...". The last remark stems from the fact that spontaneous emission occurs at a rate that would be induced by the presence of one quantum per degree of freedom whereas the zero-point oscillations provide only one half quantum. Again, Baym[3] writes: "In a sense, one can regard spontaneous emission as induced emission due to the vacuum fluctuations of the electromagnetic field".

These references are cited to show that our initial question is not an empty one and to provide the backdrop for an earlier discussion[4] in which it was shown that zero-point quanta are not needed to compute the rate of spontaneous emission.

We complete the symmetry (or anti-symmetry) of our thesis that most physicists continue after some fifty years to harbor both the concept that zero-point quanta cause spontaneous emission and a concept that is completely contradictory to the idea that spontaneous emission is caused by zero-point quanta. Here, it is sufficient to point out that few if any physicists would object to the early classical argument[5] used to approximate the rate of emission of Cerenkov radiation - another example of spontaneous emission. Indeed, the present paper treats spontaneous emission of phonons essentially classically. Other examples can readily be cited. The essential point of logic here is that if spontaneous emissions were indeed caused by zero-point quanta there would be no possibility of even remotely approximating the rate of spontaneous emission by a classical argument since the essential causal ingredient, namely, zero-point quanta, are not present in classical physics.

The causal physics is, as was shown in ref. 4, essentially classical but subject to certain obvious quantum contraints such as the quantization of energy, the finite size of an electron given by its uncertainty radius, $\hbar/mv$, and the existence of a ground state, $\frac{1}{2}\hbar\omega$.

EINSTEIN'S DETAIL BALANCE

Einstein[6] showed in 1917 that oscillators in equilibrium with a field of radiation satisfied, by an argument of detailed balance, the following simple relations:

$$\text{rate of excitation of oscillators} = \alpha n \qquad (1a)$$
$$\text{rate of emission by oscillators} = \alpha(n+1) \qquad (1b)$$

where $\underline{n}$ is the density of quanta in phase space given by the Planck distribution ($n=[\exp(\hbar\omega/kT)-1]^{-1}$ and α is a coupling constant. Equa. 1a describes induced excitation. Equa. 1b describes induced emission. The fact that induced excitation and induced emission should both be proportional to the density $\underline{n}$ of radiation is to be expected both classically and quantum mechanically. The additional term of unity in equa. 1b was called by Einstein "an effect without any apparent cause". Bohr called it "spontaneous emission". No doubt Bohr's choice of the label "spontaneous" emission contributed in no small part to the mystery or confusion subsequently attached to spontaneous emission as we have outlined in the previous section.

Our primary concerns here are that the coupling constant α is the same for both induced emission and excitation as it is for spontaneous emission and that the cause of spontaneous emission is essentially classical.

OUTLINE OF CLASSICAL APPROACH TO ELECTRON-PHONON INTERACTIONS

The fact that the coupling constant between electrons and phonons is the same for induced emission as for spontaneous emission means, logically, that if spontaneous emission can be analysed in classical terms then the coupling constant α for induced interactions can also be understood in the same terms. The approach here is to present a classical formalism for spontaneous emission. That means a formalism for computing dE/dt, the average rate of loss of energy of electrons to phonons. This classical rate is, of course, a smooth continuous rate.

The classical rate is then related to the actual rate of quantum transitions by:

$$\left.\frac{dE}{dt}\right|_{classical} = \frac{\hbar\omega}{\tau_e} \qquad (2)$$

where $\hbar\omega$ is the quantized energy of the phonons and τ_e is the "spontaneous" rate of emitting phonons. Once having determined τ_e, the total rate of interaction of electrons and phonons, that is the collision time τ_c for computing mobility ($\mu = \tau_c\ e/m$) is given by the Einstein relations as:

$$\tau_c = \frac{\tau_e}{2n+1} = \frac{\hbar\omega}{(2\ n+1)\ \frac{dE}{dt}} \qquad (3)$$

(m is the effective mass of an electron, a quantum mechanical concept).

In brief, the essence of this paper is to outline a classical procedure for computing dE/dt. It will turn out that the procedure is equally valid for amorphous as well as crystalline materials.

In an earlier series of papers(7) this formalism was discussed in more detail and was used to derive classically all of the four types of electron-phonon interactions that have appeared in the literature as derived by conventional quantum mechanical arguments. Included in these papers are also the interaction of electrons with plasmons, x-ray levels, Cerenkov radiation and the various acoustoelectric effects. All of these phenomena are characterized by the same simple, classically derived coupling constant.

DIMENSIONAL ARGUMENT FOR dE/dt

Figure 1 gives a dimensional argument for the rate of loss of energy by a moving particle to a system with which it interacts. In Fig. 1 a particle moves with velocity v past a series of elements of dimension d. The particle repels each element with a force such that, for a stationary particle, an energy E_w is stored in the "compressed spring" of the element. The frequency of vibration of each element is denoted by ω.

By inspection, we can write the maximum rate of loss of energy by the particle:

$$\left.\frac{dE}{dt}\right|_{max} = E_w\ \omega \qquad (4)$$

At the assumed velocity ωd, the particle spends a time ω^{-1} opposite each element and deflects the element almost as far as it would if the particle were stationary. While the element is in the deflected state the particle moves on to the next element leaving behind an energy of almost E_w per element. Since, by assumption, the particle traverses ω elements per second the maximum rate of energy loss is given by Equa. 4.

Returning to Fig. 1 we can write down, again, almost by inspection the rate of loss of energy for velocities greater than ωd:

$$\frac{dE}{dt} = E_w\ \left(\frac{\omega d}{v}\right)^2\ \frac{v}{d} = E_w\ d\ \frac{\omega^2}{v} \qquad (5)$$

The energy imparted to each element during the transit of the particle is proportional to the square of the momentum imparted. The momentum imparted is, in turn, proportional to the transit time d/v. Thus, the maximum momentum and energy is imparted when $d/v = \omega^{-1}$ as in Equ. 4. For higher velocities, the transit time past an element and consequently the momentum imparted decreases as v^{-1}; the energy imparted decreases as the square of the momentum, that is, as v^{-2}. Thus, the energy imparted per element is

$$E_w \left(\frac{\omega d}{v}\right)^2$$

Further, the number of elements traversed per second is v/d. The combination of these two factors yields Equ. 5.

Equ. 5 is the classical or "real space" argument for dE/dt. The particle in Fig. 1 represents the electron in the case of electron-phonon interactions and the elements of dimension d together with their attendant springs represent the medium with which the electron interacts. It remains then to express E_w in terms of the physics of interaction between an electron and its surrounding medium.

The maximum energy that an electron can impart to its surrounding medium is the coulomb energy e^2/d of the electron where d is the uncertainty radius, that is, the size $\hbar/mv$ of the electron. This maximum energy will occur when the coulomb field e/d^2 is completely polarized out by the medium. An example is an electron immersed in a medium of arbitrarily large dielectric constant. In general the energy imparted to the medium will range between zero and e^2/d. Hence, the energy imparted can be written formally as

$$E_w = \beta \frac{e^2}{d} \qquad (6)$$

where $o \leq \beta \leq 1$ is the formal coupling constant. Note that this coupling constant has a simple, physical rational scale from zero to unity and that the coupling constant is universally valid for any medium and any coupling mechanism.

In the particular case of electron-phonon interactions Equ. 6 takes the form:

$$E_w = \beta \frac{e^2}{K_H d} \qquad (7)$$

where K_H is the high frequency part of the dielectric constant ascribable to the electronic part of the medium as opposed to the ionic or atomic part of the medium. The meaning of the factor K_H is that, in general, the electron velocity is not large enough to leave behind in its trail the polarization of the electronic part of the medium. The moving electron reversibly polarizes the electronic part of the medium. Hence, the coulomb energy of the electron available for doing work on the medium and leaving that energy behind is reduced from e^2/d to $e^2/K_H d$.

We now combine Equ. 5 and 7 to obtain the general form for the rate of energy loss to phonons:

$$\frac{dE}{dt} = \beta \frac{e^2 \omega^2}{K_H v} \qquad (8)$$

Insertion of Equ. 5 into Equ. 3 now yields the value of the collison time τ_c for computing the electron mobility:

$$\tau_c = \frac{K_H \hbar v}{(2n+1) \beta e^2 \omega} \qquad (9)$$

Note that the question of whether the medium was crystalline or amorphous did not enter into the argument. Hence, Equ. 8 and 9 are equally valid for amorphous as well as crystalline materials.

THE MEANING OF β

The central physics of electron-phonon interactions lies in the coupling constant β. It was formally defined as the ratio of the energy of the polarized dielectric left behind by the moving electron to the available coulomb energy ($e^2/K_H d$) of the electron. Here d is the uncertainty radius ($\hbar/mv$) of the electron.

In an earlier publication(7) it was shown that the formal definition of β was also equal to:

$$\beta = \left. \frac{\text{Electrical Energy}}{\text{total energy}} \right|_{\text{Polarized dielectric}}$$

$$= \left. \frac{\text{Electrical Energy}}{\text{Electric \& Elastic Energy}} \right|_{\text{Polarized Dielectric}}$$

In the case of electron interaction with the polar optical phonons of an alkali halide β was evaluated as:

$$\beta = \frac{K_L - K_H}{K_L}$$

where K_L was the total or low frequency dielectric constant and K_H the electronic or high frequency part of the dielectric constant. For electron energy loss to plasmons β = 1 as expected from the lack of elastic energy in the polarized plasma.

CALCULATION OF dE/dt

In the earlier publication(7) the present classical model of spontaneous emission was used to derive the rates of spontaneous emission of some seven phenomena listed in Table I. In all cases the published results obtained by the more orthodox quantum mechanical derivations were matched except for a small numerical constant ascribable to the approximations used in our derivations. The bracketed factors in Table I are the expressions for our coupling constantants β. The quantum constraints are listed in each case in the Table.

SELF TRAPPED ELECTRONS

Considerable literature has been devoted to the question of whether an electron interacting with, for example, acoustic phonons in a co-valent solid can be self trapped so that it must migrate by a temperature activated hopping process. In particular, Toyazawa(16) computed the criterion for self trapping in a covalent solid. The criterion involved the deformation potential of the solid. His choice of deformation potential to satisfy his criterion for self trapping was large enough to violate our logical constraint that the constant β must logically be less than unity. Hence, one must conclude that self trapping cannot occur(7).

Self trapping can occur for reasons other than self trapping by phonons. A likely source is the irregularities in the bottom of the conduction band in disordered solids. Also electronic effects as opposed to phonon effects can lead to self-trapping. The presence of impurities is an obvious example. It is also possible that in the case of materials like sulphur that have large atomic electron affinity that electronic effects may combine with phonon effects to make self trapping possible.

Table 1. Time Rates of Energy Loss (dE/dt) by Electrons of Velocity v.

Phenomenon	dE/dt	Source
Polar Optical Phonons	$\left[\frac{\epsilon_0 - \epsilon_\infty}{\epsilon_0}\right] \frac{e^2\omega^2}{\epsilon_\infty v} \ln\left(\frac{2mv^2}{\hbar\omega}\right)$ Note: $\frac{1}{2}mv^2 > \hbar\omega$	Fröhlich[8] Callen[9]
Piezoelectric Phonons	$\frac{\pi}{4}\left[\frac{\epsilon_p^2}{KC}\right] \frac{e^2\omega^2}{Kv}$ Note: $2mv = \hbar\omega/v.$	Tsu[10]
Acoustic Phonons	$\frac{3}{4}\left[\frac{B^2\omega^2 K}{4\pi e^2 \rho v_s^4}\right] \frac{e^2\omega^2}{Kv}$ Note: $2mv = \hbar\omega/v.$	Seitz[11] Conwell[12]
Nonpolar Optical Phonons	$\frac{1}{2}\left[\frac{\pi K D^2}{\rho e^2\omega^2\lambda^2}\right] \frac{e^2\omega^2}{Kv}$ Note: $\frac{1}{2}mv^2 > \hbar\omega$ and $\frac{\lambda}{2\pi} = \frac{\hbar}{2mv}$	Conwell[12]
X-Ray Levels	$\left[\frac{\omega_p^2}{\omega_e^2}\right] \frac{e^2\omega_p^2}{v} \ln\left(\frac{2mv^2}{\hbar\omega_e}\right)$ Note: $\omega_p^2 = \frac{4\pi ne^2}{m} \ll \omega_e^2$ $\hbar\omega_e$ = Excitation energy of x-ray levels $\frac{1}{2}mv^2 > \hbar\omega_e$	Bohr[13] Bethe[14]
Plasma	$\left[1\right] \frac{e^2\omega^2}{v} \ln\left(\frac{2mv^2}{\hbar\omega}\right)$ Note: ω =-plasma frequency and $\frac{1}{2}mv^2 > \hbar\omega$	Bohm and Pines[15]
Cerenkov	$\left[1 - \frac{1}{\epsilon_r}\right] \frac{e^2\omega^2}{v}$ Note: $v = c$ = velocity of light in vacuum	See, e.g., Schiff[2]

Definitions for Table 1

ϵ_0 = low-frequency dielectric constant
ϵ_∞ = high-frequency (optical) dielectric constant
K = dielectric constant
ϵ_p = piezoelectric constant
C = elastic modulus (dynes/cm^2)
ρ = density (grams/cm^3)
v_s = phase velocity of sound
v = velocity of electron
ω = angular frequency of radiation
B = deformation potential (electron volts in ergs/unit strain)
D = optical deformation potential (electron volts in ergs per centimeter relative shift of sublattices)
m = effective mass of electrons

MAGNITUDE OF MOBILITY IN AMORPHOUS MATERIALS

The import of the present analysis of mobility in amorphous materials is that the contribution to mobility of the inelastic scattering by electron-phonon interactions should, in general, be comparable for amorphous and crystalline materials. This statement is based on the fact that the measure of this interaction is the coupling constant β, that is, the polarization of the solid (amorphous or crystalline) by the electron. The polarization of the medium by the electron should not, in general, be sensitive to whether the medium is amorphous or crystalline.

The measured mobilities in amorphous materials range from near unity to 400 for liquid methane and 2,200 for liquid Xenon.[17] These large values are evidence that the amorphous state does not necessarily lead to low mobilities. Some of the very low observed values are very likely due to non-phonon effects such as elastic scattering from the types of disorder that lead to irregularities in the bottom of the conduction band. Measurements of the thermalization rate of electrons excited to a few tenths of a volt above the bottom of the conduction band[18] of amorphous silicon can be interpreted as yielding relatively high mobilities for the phonon contribution to mobility. Similarly, the remarkably low values for geminate recombination in amorphous silicon are strong evidence for relatively large mean free paths and, hence, for relatively high phonon mediated mobilities.

SUMMARY REMARKS

The present treatment of mobility in essentially classical terms yields not only a relatively simple formalism and one that emphasizes a physical picture in real space, but also contains a number of points of broader significance. These are:

1. The cause of spontaneous emission is the classical interaction between an electron and its surrounding medium rather than the widely cited zero-point quanta.

2. The phonon component of mobility is derived from the rate of spontaneous emission and is closely the same for amorphous and crystalline materials.

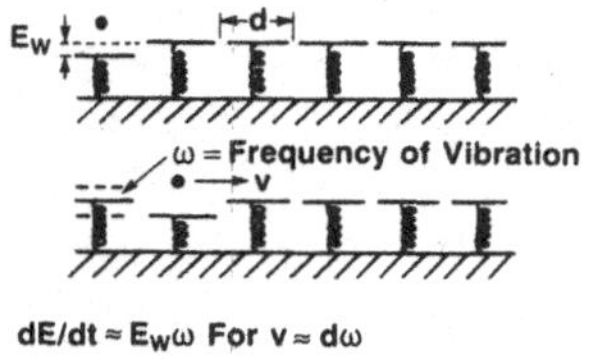

$$dE/dt \approx E_W \omega \text{ For } v \approx d\omega$$

$$dE/dt \approx E_W (d\omega/v)^2 \, v/d \text{ for } v >> d\omega$$

$$= (E_W d)\, \omega^2/v$$

Figure 1. Model for Computing rate of energy loss by a moving particle

3. A simple, rational and universal coupling constant β ($0 < \beta < 1$) is defined.

4. The same β is valid the quantum phenomena of electron-phonon interactions and for the classical acousto-electric effects.

5. The criterion for self-trapping published by Toyazawa leads to values of
β greater than unity and is, therefore, logically invalid.

References

1. R. Eisberg and R. Resnick, Quantum Physics, John Wiley, Inc. New York,
1974 (p. 316).
2. L. I. Schiff, Quantum Mechanics McGraw-Hill Book Co., New York, 1955.
3. G. Baym, Lectures on Quantum Mechanics W. A. Benjamin, Inc. Reading, MA, 1973 (p. 276).
4. A. Rose Phys. Stat. Sol (a) 61, 133 (1980).
5. I. Frank and I. Tamm Comptes Rendus (Doklady) de l'Acad. des Sci., USSR 14, 109 (1937).
6. A. Einstein, Phys. Z. 18, 121 (1917).
7. A. Rose RCA Review 27, 98 (1966); 27, 600 (1966); 28, 634 (1967); 30, 435 (1969); 32, 463 (1971).
8. H. Frohlich Proc. Roy. Soc. (London) A188, 521 (1947) and A160, 230 (1937).
9. H. B. Callen Phys. Rev. 76, 1394 (1949).
10. R. Tsu J. Appl. Phys. 35, 125 (1964).
11. F. Seitz Phys. Rev. 76, 1376 (1949).
12. E. M. Conwell Phys. Rev. 135, A1138 (1964).
13. N. Bohr Phil. Mag. 25, 10 (1913) and 30, 581 (1915).
14. H. Bethe Ann. Phys. 16, 285 (1933).
See also J. D. Jackson, Classical Electrodynamics John Wiley and Sons, New York, 1963.
15. D. Pines and D. Bohm Phys. Rev. 85, 338 (1952) and 92, 609 (1953).
16. Y. Toyazawa Polarons and Excitons p. 211 Oliver and Boyd, London (1963)
ed. by C. G. Kuper and G. D. Whitfield.
17. Werner F. Schmidt, IEEE Trans. on Electrical Insulation EI-19, 389 (1984).
18. Z. Vardeny and J. Tauc Phys. Rev. Lett. 46, 1223 (1981).

Proc. of the Int. Workshop on Amorphous Semiconductors, pp. 29–34
edited by H. Fritzsche, D.-X. Han & C.C. Tsai

A SIMPLE CLASSICAL APPROACH TO MOBILITY IN AMORPHOUS MATERIALS

Albert Rose

Visiting Scientist
Chronar Corp.
Princeton, N.J. 08540, U.S.A.

ABSTRACT

A simple and essentially classical formalism valid for both amorphous and crystalline materials is used to compute the phonon component of electron mobilities in amorphous materials. A significant part of this formalism is a simple physical concept of a coupling constant valid for all of the electron interactions in a solid both classical and quantum.

1. OVERVIEW

The conventional treatment of electron-phonon interactions is to compute the rate of transition of an electron wave function from one state to a neighboring state under the perturbation of a phonon. In the process a phonon is emitted or absorbed and the electron is scattered. The time to emit (or absorb) a phonon whose momentum is equal to the electron momentum is the mean free time τ in the expression for mobility $\tau \frac{e}{m}$. The mean distance

30

traveled by the electron between phonon collisions is the mena free path. This calculation in terms of wave functions fails to present the problem in terms of the elementary classical concepts involved. To do this, the electron should be represented as a particle rather than a wave. The size of the particle is $\hbar/mv$. Given the electron as a particle, its rate of loss of energy to the lattice can be computed quite simply and generally as a Cerenkov type of radiation. If the rate of loss of energy is dE/dt the time to emit a phonon of energy $\hbar\omega$ is given by:

$$\frac{dE}{dt} = \frac{\hbar\omega}{\tau_e} \tag{1}$$

where τ_e is the spontaneous rate of emitting phonons. According to Einstein's[1] detail balance argument the rate of interacting with phonons is

$$\tau_c = \tau_e\ (2n + 1)^{-1} \tag{2}$$

where τ_c is the mobility scattering time so that

$$\mu = \tau_c \frac{e}{m} = \tau_e \frac{e}{(2n + 1)\ m} \tag{3}$$

where n is the Planck density of phonons

$$n = \exp\left[\frac{\hbar\omega}{kT} - 1\right]^{-1}$$

The electron in the solid polarizes the solid and forms an energy well E_W into which the electron sinks. When the electron moves slowly it carries the energy well with it without loss of energy, much as a ball on a rubber membrane. When the electron moves faster than the velocity of sound (phonon velocity) it leaves in its trail the energy of the energy well. The maximum rate of loss of energy is

$$\left.\frac{dE}{dt}\right|_{max} = E\ \omega \tag{4}$$

where ω is the phonon frequency. The argument is that an electron inserted in a solid will form the energy well E_W in a time ω^{-1}. Let the size of the energy well $d = \frac{\hbar}{mv}$. Then, if the electron is suddenly moved a distance d from its initial position it will leave its energy well behind as radiated phonon energy and will form a new energy well in a time ω^{-1}. This process can be repeated times per second and will correspond to an average electron velocity of ωd. Also the average rate of loss of energy will be $E_W \omega$. When the electron moves with a velocity v greater than ωd the rate of loss of energy is

$$\frac{dE}{dT} = E_W \left(\frac{\omega d}{v}\right)^2 \frac{v}{d} = E_W d \frac{\omega^2}{v} \qquad (5)$$

The factor $\left(\frac{\omega d}{v}\right)^2$ reflects the fact that the energy imparted to the medium is proportional to the square of the momentum imparted to the medium. Also the momentum imparted decreases as v^{-1}. The result is that the energy well is reduced by the factor $\left(\frac{\omega d}{v}\right)^2$. The number of energy wells formed per second is v/d. These two effects yield equ. 6 as the general expression for the rate of loss of energy for velocities greater than ωd. Note that equ. 5 gives the rate of loss of energy for spontaneous emission. It does so without resort to zero point quanta as the cause of spontaneous as is commonly cited in the literature.

GENERAL FORMALISM FOR MOBILITY

The energy well E_W formed in the medium by the electron comes at the expense of the coulomb energy of the electron. For this reason the maximum value of E_W is the available Coulomb energy $\frac{e^2}{K_H d}$ where d is the uncertainty radius $\hbar/mv$ of the electron and K_H is the high frequency or electronic part of the dielectric constant. When the moving electron polarizes the medium, it polarizes both the electronic and ionic components. Only the ionic polarization is left behind since the electronic part is reversibly polarized. Hence, the value of E_W can be written

$$E_W = \beta \frac{e^2}{K_H d}, \quad 0 \leq \beta \leq 1. \tag{6}$$

Combination of equs. 5 and 6 then yields the generalized form for rate of energy loss:

$$\frac{dE}{dT} = \beta \frac{e^2}{K_H} \frac{\omega^2}{v} \tag{7}$$

Combination of equs. 1,3 and 7 yields the generalized form for mobility:

$$\mu = \tau_e \frac{e}{(2n+1)m}$$

$$= \frac{\hbar\omega}{dE/dt} \frac{e}{(2n+1)m}$$

$$= \frac{\hbar \quad K_H \quad v}{\beta \; e^2 \omega \quad (2n+1)} \tag{8}$$

In an earlier publication[2] it was shown that the formal definition of β was also equal to:

$$\beta = \left.\frac{\text{Electrical Energy}}{\text{Total Energy}}\right|_{\text{polarized dielectric}}$$

$$= \left.\frac{\text{Electrical Energy}}{\text{Electric plus Elastic Energy}}\right|_{\text{polarized dielectric}} \tag{9}$$

In this publication[2] the electron interaction with polar optical phonons of an alkali halide, β was evaluated as

$$\beta = \frac{K_L - K_H}{K_L} \tag{10}$$

where K_L was the total or low frequency dielectric constant and K_H the electronic or high frequency part of the dielectric constant. It is clear from equs. 9 and 10 that β must logically be ≤ 1. This fact plays a significant role in the problem of self trapped electrons. For example, Toyazawa[3] computed the criterion for electron self trapping in a co-valent solid. His choice of deformation potential for self trapping turns out to violate the logical constraint that β must be less than unity.

DISCUSSION

The present treatment of mobility in essentially classical terms yields not only a relatively simple formalism and one that emphasizes a physical picture in real space, but also contains a number of points of broader significance.

(1) The major point is that the coupling constant β has a logical upper limit of unity. For this reason it rules out self trapping of the electron because Toyazawa's criterion for self trapping leads to a value of β greater than unity.

(2) The cause of spontaneous emission is the classical interaction between an electron and its surrounding medium rather than the widely held zero point quanta, a quantum mechanical phenomenon.

(3) The phonon component of mobility is derived from the rate of spontaneous emission and is closely the same for amorphous and crystalline materials (see equ. 10).

In support of this statement, measured mobilities of 400 cm^2/volt sec. have been reported by Schmidt[4] for liquid methane and 2,200 cm^2/vol sec. for liquid xenon. Also the rates of thermalization in amorphous silicon reported by Vardeny and Tauc[4] are consistent with high values of the phonon component of mobility.

REFERENCES

1. Einstein, A. Phys. Z. 18, 12 (1917)

2. Rose, A. RCA Review 27, 98 (1966); 27, 600

(1966); 28, 634(1967); 30, 435(1969); 32, 463 (1971)

3. Toyazawa, Y., Polarons and Excitons, p. 211, Oliver and Boyd, London (1963), e.d. by C.G. Kuper and G.D. Whitfield
4. Schmidt, W.F., IEEE Trans. on Electrical Insulation EI-19, 389(1984)
5. Vardeny, Z. and Tauc, J., Phys. Rev. Lett. 46, 1223(1981).

AN EXTENSION OF EINSTEIN'S TREATMENT OF SPONTANEOUS EMISSION

Albert Rose

Visiting Scientist
Exxon Laboratories, Clinton, NJ and
Chronar Laboratories, Princeton, NJ

ABSTRACT

Einstein's detailed balance argument in which the concept of spontaneous emission was first introduced is used to demonstrate that spontaneous emission is a classical phenomenon and is not a consequence of zero point quanta as is widely believed. The average rates of photoinduced excitation and emission are also computed classically.

The concept of spontaneous emission is universally recognized as having been introduced by Einstein using a detailed balance argument between a gas of atoms in thermal equilibrium with an ambiance of photons.[1] For the past seventy years a more or less steady stream of theoretical papers has addressed the question of what causes spontaneous emissions.[2]

The detailed balance was expressed as:

$$n_2 \quad (A + B\rho) = n_1 \quad B\rho \qquad (1)$$

where n_2 is the density of excited electrons, n_1 is the density of electrons in the ground state, A is the spontaneous rate emission of photons, $B\rho$ is the induced rate of emission of photons, and also is the induced rate of absorption of photons, and ρ is the density of photons whose energy ΔE was equal to the energy difference between the excited and ground states. Einstein showed independently that B was the same for optically induced emission and absorption. The ratio of n_1 to n_2 followed Boltzman's law

$$\frac{n_1}{n_2} = \exp\left(\frac{\Delta E}{k\,T}\right) \qquad (2)$$

From Eqs. (1) and (2) Einstein deduced Planck's radiation law:

$$\rho = \frac{A/B}{\exp\left(\frac{\Delta E}{k\,T}\right) - 1} \qquad (3)$$

In place of Eq. (3) we solve for:

$$B\rho = \frac{A}{\exp\left(\frac{\Delta E}{k\,T}\right) - 1} = A\,n_p \qquad (4)$$

where $n_p = \dfrac{1}{\exp\left(\frac{\Delta E}{k\,T}\right) - 1}$

= The Planck density of photons per cell in phase space.

Insertion of Eq. 4 into Eq. 1 then yields

$$n_2\,(A + A\,n_p) = n_1\,A\,n_p \qquad (5)$$

From Eq. 5 we deduce three major conclusions which are contained implicitly in Einstein's Eq. 1 and which are not explicitly discussed in the literature.

(1) The explicit value of A is:

$$A = \frac{e^2\,a^2}{3\;c^3}\left(\frac{8\pi\,\nu^3}{c^3}\right) \qquad (6)$$

where a is the acceleration of the electron in its excited orbit and A is the spontaneous rate of loss of energy by an excited electron. $\frac{8\,\pi\,\nu^3}{c^3}$ is the density of cells in phase space.

Einstein states in his paper that "It is well known that a vibrating Planck resonator emits, according to Hertz, energy independent of whether it is excited by an external field or not".

The only possible interpretation is that Einstein believed the cause of spontaneous emission to be the well known classical phenomenon of radiation by an accelerated electron. It is at least puzzling why Einstein did not cite Eq. 6 in his paper.

(2) The cause of spontaneous emission is a classical phe-

nomenon as opposed to the widely held view that spontaneous emission is a quantum mechanical effect due to zero point quanta.[2]

(3) Since, in Eq. 5, the coefficients of induced emission and optical absorption are the same as the coefficient of spontaneous emission, it must be concluded that the processes of induced emission and optical absorption are classical phenomena.

Consistent with item 3 Heitler[3] states that his classical derivation of the rate of spontaneous emission is "almost identical" with his derivation of that rate from quantum electro dynamics. That rate is

$$\frac{dE}{dt} = \frac{16\pi^4 e^2 \nu^4 X^2}{3c^3} \qquad (7)$$

and is readily derived from Eq. 6. since X is the amplitude of the vibrating electron, and $a^2 = 16\pi^4 \nu^4 X^2$.

We have argued that the rates of induced emission and absorption must be classically derived rates since in Eq. 5 the coefficient A is the same for induced and spontaneous emission and since by Eq. 6 the rate of spontaneous emission is a classically derived rate. We show now that the rates of induced transitions can also be independently derived from the elementary classical expression:

$$\frac{dE}{dt} = \epsilon e v \qquad (8)$$

Here, ϵ is the electric field of the light wave and v is the drift velocity of the vibrating electron that vibrates at the frequency ν.
Hence,

$$v = \epsilon\mu = \epsilon\frac{\tau e}{m} \qquad (9)$$

We take τ to be the time between field reversals, namely, ν^{-1}. The field reversal plays the same role as electron-phonon collisions in the normal expression $\epsilon\frac{\tau e}{m}$ for drift velocity. That is, the field reversal reduces the forward velocity of the electron to zero as does the electron-phonon collision. The difference between the two is that the phonon-emission is an inelastic collision in which the energy of the electron is dissipated to the lattice whereas the field reversal constitutes an elastic collision in which the energy imparted to the electron is conserved. In this way the work done by the field on the electron is accumulated during successive collisions. In the normal case of drift velocity, $\epsilon\mu$, if the lattice collisions are assumed to be elastic instead of inelastic, the accumulated energy appears as the kinetic energy of a hot electron.

Combination of Eqs. 8 and 9 yields:

$$\frac{dE}{dt} = \epsilon^2 e^2 \frac{\tau}{m} \tag{10}$$

Since the energy of the photon is quantized, the accumulated energy during absorption (or the radiated energy during emission) is $h\nu$. Hence,

$$h\nu = \frac{\epsilon^2 \lambda^3}{8\pi} = 1/2 \; mv^2 \tag{11}$$

or

$$\epsilon^2 = 4\pi \, m \frac{\nu^3}{c^3} v^2 \tag{12}$$

From Eqs. 10 and 12:

$$\frac{dE}{dt} = 4\pi \frac{\nu^3}{c^3} e^2 \tau v^2$$

$$= \frac{16 \pi^3 e^2 X^2 \nu^4}{c^3} \tag{13}$$

where $v = 2\pi X\nu$

$\tau = \nu^{-1}$

Since eq. 13 matches (except for the minor approximation $\pi = 3$) eq. 7, Heitler's result derived from quantum electrodynamics, we have shown that the classical result equ. 13 is essentially the same as the quantum result eq. 7.

The quantity $\frac{dE}{dt}$ should be regarded, both classically and quantum mechanically, as the average value of a stochastic process. The classical solution should not be used to compute the detailed path of the electron between the end points of its jump. Such an interpretation would violate the uncertainty principle. Also, while equ. 13 is the result of a classical calculation for $\frac{dE}{dt}$, the end points of the jumps are determined by quantum mechanics. The quantum jumps treated here are confined to allowed dipole transitions.

On the face of it, it is remarkable that the three expressions for the rate of energy change per electron and per photon for induced emission and absorption and per electron for spontaneous emission are identical. Einstein's detailed balance requires this identity whatever the physics involved. The three expressions are:

$$\frac{e^2 a^2}{3 c^3}$$

ϵev

and

$$\frac{16 \pi^4 e^2 \nu^4 x^2}{3 c^3}$$

and encompass the radically different physics of radiation by an accelerated electron and energy transfer (ϵev) by the simplest mechanism of force times velocity.

In summary, Einstein's treatment of spontaneous emission has been extended to show that the average rates of optical absorption and emission can be derived classically. Moreover, there is no need to introduce the model of zero point quanta to account for spontaneous emission as has been done frequently over the last seventy years. Hence, there is no need to try to reconcile the [½] $h\nu$ of zero point quanta with the required value of [1] $h\nu$.4

A final point of novelty is the translation of optically induced emission and absorption into the simplest of classical expressions:

$$\frac{dE}{dt} = \epsilon ev$$

We conclude by calling attention to the fact that Einstein's logic is equally valid for the spontaneous emission of phonons as for photons. Hence, if zero point quanta are used as the cause of spontaneous emission of photons they should also be used as the cause of spontaneous emission of phonons. In reference 5 we have derived the rate of emission of phonons by classical arguments for some five types of electron-phonon coupling. The results closely approximate the conventional derivations in the literature using quantum mechanics. What we wish to point out is that none of the literature citations in this reference chose to make use of zero point quanta as the cause of spontaneous emission nor to my knowledge has that connection ever been used.

Acknowledgements

I am indebted to George Cody and Ping Sheng of Exxon Laboratories, Rober Dicke of Princeton University, Freeman Dyson of the Institute for Advanced Study, and Richard Williams of RCA Laboratories for helpful discussions.

References

1. A. Einstein, Phys. Z. 18, 121 (1917).

2. R.R. Puri, J.O. S.A. 2, 447 (1985).
   B. Fain, IL Nuovo Cimento 63, 73 (1982).

3. W. Heitler, The Quantum Theory of Radiation, p. 128, The University Press, Oxford, 1953.

4. Leonard Schiff, Quantum Mechanics, p. 400, McGraw-Hill, New York, 1955.

5. A. Rose, Physics of Disordered Materials, p. 391 Edited by D. Adler, H. Fritzsche & S.R. Ovshinsky, The Plenum Press, New York, 1985.